SEDIMENT IN STREAMS

Support for the preparation of this book was provided by

Duke Power Company
Charlotte, North Carolina

Support for the publication of this book was provided by

Duke Power Company
Charlotte, North Carolina

U.S. Environmental Protection Agency
Washington, DC

U.S. Department of the Interior
Washington, DC
Bureau of Reclamation
Bureau of Land Management

SEDIMENT IN STREAMS

Sources, Biological Effects, and Control

Thomas F. Waters
Department of Fisheries and Wildlife
University of Minnesota
St. Paul, Minnesota

American Fisheries Society Monograph 7

American Fisheries Society
Bethesda, Maryland
1995

The American Fisheries Society Monograph series is a registered serial.
A suggested citation format for this monograph is

Waters, T. F. 1995. Sediment in streams: sources, biological effects, and control.
American Fisheries Society Monograph 7.

Photographs by the author except where noted

Library of Congress Catalog Card Number: 95-079171

ISBN 0-913235-97-0   ISSN 0362-1715

Printed in the United States of America on recycled, acid-free paper

Published by the American Fisheries Society

To Carol

CONTENTS

PREFACE

Probably no stream fisheries biologist or stream ecologist has been involved in a research project for any length of time without serious sedimentation occurring to upset the research plan. Such was the case in my own experience, several times. In Valley Creek, Minnesota, just at the start of a long-term trout production study, a heavy rainstorm one night in 1965 resulted in sand covering the entire streambed, nearly eradicating the fish population. Several years later, in the middle of the study, poor soil-handling operations in a housing project in the upper watershed resulted in clay sedimentation that decimated both trout and invertebrate populations for 2 years. Still later, after heavy sedimentation, I visited some farmland suspected as a sediment source, and there I saw plowed furrows that led downslope straight toward the head of a large, eroding gully.

Such events were infuriating, but 30 years ago little concern about sediment was evident among resource agencies or my colleagues in fisheries and stream ecology.

So, when in 1990 Kenneth Manuel and David Braatz, biologists in the environmental laboratory of Duke Power Company, Charlotte, North Carolina, suggested that I conduct a literature review in the subject of stream sediment, I was quickly convinced of its need, and in the spring of that year the three of us hatched the project. At the time, such a review seemed to me to be a relatively minor task that I could complete outside of a full schedule related to my approaching retirement from the University of Minnesota.

I was soon disabused of this notion. It was readily apparent that an immense literature had been published. This literature already included a number of reviews, although these were all limited in one way or another in scope. For 30 years there has been no summary published that was comprehensive in terms of geographic coverage, completeness through major taxonomic groups in streams, or in relation to sediment sources. I immediately found the review by Cordone and Kelley (1961), which although comprehensive in scope, included few references—reflecting the primitive state of knowledge of the subject at that time. However, their paper was seminal; for many years almost every publication on sediment cited their review, and no other appeared to supersede Cordone and Kelley with comprehensive coverage.

Agency reports and other nonjournal literature made up a large proportion of the reported research material. Of particular importance were the several series of publications by the U.S. Forest Service experiment stations.

At about the same time this project was initiated, it became apparent that awareness and activity in the general area of sediment problems was increasing rapidly—in national seminars and workshops, surveys and reports of the U.S. Environmental Protection Agency, efforts of the American Fisheries Society, and a notable increase of reports in the U.S. Forest Service publication series. The timing for a comprehensive review seemed particularly auspicious.

This review is prepared chiefly from the perspective of freshwater fisheries in streams. A very large portion of the literature on stream sediment has originated in natural-resource government agencies directly or indirectly concerned with fisheries management. Another large portion of the literature has originated in pollution control agencies, whose purview includes resource management for healthy biological communities in streams. Thus, the two viewpoints—applied, for the stream's fishery, and basic, for the stream's biotic integrity—are joined in this review as a study of community-wide problems caused by anthropogenic sediment. Intensive literature searches were continued through 1992, with the goal of being as comprehensive as I deemed beneficial, through that date. Additional reviews, books, and major critical papers that subsequently reached my attention also have been included.

Many sediment effects are species-specific with respect to fishes, but only common names of fishes are included in this review. In order to avoid the distractions of inserted Latin names, these are listed for convenience in the front of the book. Much of the sediment-related literature concerns trout and salmon, members of the family Salmonidae, subfamily Salmoninae. Although a more formal usage today would be *salmonine*, the common name *salmonid* still enjoys the most common usage for trout and salmon. This older term, *salmonid*, is used throughout this volume, and it is hoped that the reader will realize that fishes other than trout and salmon are not intended.

The problem of units was particularly difficult, because disciplines are not consistent in that regard. Changing all units to either English or metric, I believe, would lose the degree of precision, intended or not intended, of many of the original authors' uses. Consequently, I have retained the units originally used by the authors, either English or metric, and I have included a table in the front of the book for convenience in making desired conversions.

With a single exception, graduate theses and dissertations have not been included in this review. Those theses cited in the published literature (i.e., those presumably having the most recognized significance) were almost invariably followed by a publication in a major journal. However, theses and dissertations that I noted as cited in published papers are listed in Appendix IV.

Definitions that varied among disciplines and other limitations also needed to be dealt with. Even the term "sediment" has different meanings between perspectives. I quickly encountered multiple views and interests and needed to sharply define my own scope and objectives. Consequently, the following six topics, tangentially related to the biological effects of inorganic sediment, are not addressed in this volume; they are all recognized as distinctive research topics with large literatures of their own:

1. Hydrology–Sedimentology. An immense body of literature on physical processes in streams is included in fluvial geomorphology and related disciplines. The first major treatise was the classic *Fluvial Processes in Geomorphology*, by Leopold, Wolman, and Miller (1964). Recent significant publications include the reference books by Hey et al. (1982), Richards (1982), Thorne et al. (1987), Black (1991), Maidment (1993), and the most recent review of fluvial geomorphology by Rhoads (1994). The text by Gordon et al. (1992), addressing ecological problems related to hydrology, will be especially useful to biologists. The new book, *A View of the River* (Leopold 1994) is small but very readable by nonfluvial geomorphologists. Leopold describes the book as a primer—with a minimum of mathematics and almost no reference citations in the text, intended for a broad audience. "In this respect," writes the author, "it is a view of the river, the scene experienced by one individual." It will be greatly welcomed by fisheries biologists and stream ecologists.

2. Basic ecology of the stream substratum as an environment of benthic organisms, the foundation of the discipline of benthology (Hynes 1970; Barnes and Minshall 1983; Minshall 1984, 1988; Resh and Rosenberg 1984).

3. Dissolved solids, including dissolved organic matter (Lock et al. 1977; Sorensen et al. 1977; Meyer et al. 1988a). Dissolved organic matter includes organic molecules that sometimes impart color to stream water (e.g., the brownish bog color of northern, softwater streams). Dissolved organic matter constitutes an important part of the subject of allochthony, especially when particles flocculate under turbulent current conditions to form larger particles (Lush and Hynes 1973), which are then available as nutrition to other organisms. Dissolved inorganic matter includes carbonates, silicates, and other nutrients such as phosphates and nitrates, basic to primary production. A large body of literature exists on the subject of elemental and nutrient production and transport in streams (Elwood et al. 1983; Meyer et al. 1988b).

4. Organic material such as woody debris, for example, waste from logging operations (Anderson et al. 1978; Harmon et al. 1986). Other allochthonous detritus includes deciduous leaves that serve as energy sources for stream invertebrates. These resources and their processing constitute the major interest in allochthony, for which a large distinct literature exists (Cummins 1974; Anderson and Sedell 1979; Cummins and Klug 1979; Wallace and Merritt 1980).

5. Contaminated sediments, either inorganic or organic, which when transported through a stream may cause mortality or sublethal effects of several kinds on all components of the biological community. This subject is more closely allied with the greater literature on toxic water pollutants. Recent reviews and discussions include those by Baker (1980a, 1980b), Olsen and Adams (1984), and Baudo et al. (1990).

6. The large river as a specific habitat, increasing in its recognized importance. It is the relatively small stream or river that is covered in this review. Whereas this definition by itself is vague, it clearly excludes those large rivers that have their streambed almost totally covered with naturally deposited silt and sand and which may be permanently turbid, that is, the lowermost reaches in the River Continuum (Vannote et al. 1980). The emphasis of most sediment research has been on naturally clear, small streams that potentially or actually are affected

by either suspended or deposited sediment. A recent review of turbidity and suspended sediments in large rivers includes over 100 annotated references and a bibliography of 200 additional nonannotated entries (Vohs et al. 1993), although most of these concern stream problems in general and are included in the references cited in this volume. Furthermore, it would appear that control measures for sediment in large rivers must necessarily begin in upstream reaches; some reports in the literature obviously consider such measures as intended to protect larger reaches farther downstream.

Anthropogenic sediments have influences other than those that affect stream biological components. These problems have also stimulated much research and produced substantial bodies of literature. Topics include: filling reservoirs and loss of capacity, filling harbors and estuaries and effects on navigation, losing or reducing quality of domestic and industrial water supply, diminishing river aesthetics, and human health problems. Each of these topics deserves special and independent review and analysis.

The summary at the end of this book is intended to provide a condensation of the principal results derived from the literature review. It also includes some of the author's interpretations and critical research needs emerging from the review, but no reference citations. Thus, the summary gives a rapid overview of the subject, which may be sufficient for many readers. Detailed research results and documentation can all be consulted with further reading in the main body of the book.

ACKNOWLEDGMENTS

Major credit for initiating this sediment review goes to Kenneth Manuel and David Braatz of Duke Power Company, Charlotte, North Carolina. But little did I realize the great amount of additional assistance that I would eventually require to complete the immensely larger task of producing a book. Through their efforts and others, a grant was provided by Duke Power to the Department of Fisheries and Wildlife, University of Minnesota, for assistance in preparing this review.

Many individuals contributed to the search, identification, and acquisition of literature, particularly informal agency and industry reports (the "gray" literature), which are often absent from library catalogs and computer databases. Principal among these persons was John Kenney, a doctoral graduate student in my department who was instrumental in obtaining literature from around North America, and who assisted in production of the figures. The excellent reproductions of photographs included in this book were the product of the photographic skill of Malcolm B. MacFarlane, of Art Design, Eagan, Minnesota. Photographs in addition to my own were kindly provided by Don Chapman Consultants, William S. Platts, and Lloyd W. Swift, Jr. The staff of the Entomology, Fisheries, and Wildlife Library, Loralee Kerr and Sue Stegmeir, as usual with all my writing projects, gave me critical assistance with enthusiasm and efficiency in many ways. My colleague on the university fisheries faculty Raymond M. Newman and department staff assistant Shannon Flynn provided me with essential technical aid in computerized text handling. I appreciate the time spent in the field by Tex Hawkins, U.S. Fish and Wildlife Service, and Larry Gates, Minnesota Department of Natural Resources, acquainting me with sediment problems and control measures in southern Minnesota.

A host of persons provided leads, bibliographic lists, or individual reprints and copies of documents in the subject of sediment—often material that I would have missed otherwise. These include: Laurence Angle, Robert L. Delaney, and Pam Thiel, U.S. Fish and Wildlife Service; Jack Arthur and Gary W. Kohlhepp, U.S. Environmental Protection Agency; David A. Braatz, Larry Olmsted, and Robert Siler, Duke Power Company; Steve Dillard and John C. Morse, Clemson University; Henry G. Drewes, Jack Enblom, Craig Milewski, and Rick Walsh, Minnesota Department of Natural Resources; Steve Haar, U.S. Soil Conservation Service; Bret C. Harvey, Weber State University; Mark Hove, University of Minnesota; James R. Karr, University of Washington; Jacqueline D. LaPerriere and James B. Reynolds, University of Alaska; A. Dennis Lemly, Wake Forest University; Allen M. Peterson, New York State Electric & Gas Corporation; Brad Phipps, Western Kentucky University; John Quinn, New Zealand Water Quality Centre; William Turner, Missouri Department of Conservation; and Troy Zorn, Michigan Department of Natural Resources.

During these past 5 years I continuously benefitted from the assistance and support of a group of biologists and administrators in Duke Power Company. Kenneth Manuel handled financial affairs and administrative contacts, and David Braatz served as my biological contact, particularly with regional sediment projects and problems, so important in the southern Appalachian region. Others in the company met with me periodically to offer suggestions or to share their field experience with me, including Hugh Barwick, Tommy Bowen, Todd Folsom, Arnie Gnilka, Larry Olmsted, Robert Siler, and Tom Wilda. I enjoyed many field trips with Dave Braatz and others into the byways and skyways of the exquisite Blue Ridge Mountains, to view some of the sediment problems and, as well, some of the control programs in place, in that most sensitive of natural environments.

Duke Power Company, the U.S. Bureau of Land Management, the U.S. Bureau of Reclamation, and the U.S. Environmental Protection Agency all contributed financial support for the publication of the final product. The editors and staff of the American Fisheries Society have made the publication process and book production both enjoyable and professionally gratifying.

Most importantly from a personal standpoint, have been the love and loyalty from my wife, Carol, who withstood my absences, both physical and mental, during this most intense and time-consuming project.

I am grateful to all.

COMMON AND SCIENTIFIC NAMES OF FISH SPECIES

Throughout this monograph, fish species are referred to by their common names in nontaxonomic contexts. Their respective scientific names follow.[a]

Campostoma anomalum .... central stoneroller
Semotilus atromaculatus .... creek chub
Catostomus commersoni .... white sucker
Moxostoma erythrurum .... golden redhorse
Oncorhynchus clarki .... cutthroat trout
Oncorhynchus keta .... chum salmon
Oncorhynchus kisutch .... coho salmon
Oncorhynchus mykiss .... rainbow trout (steelhead trout)
Oncorhynchus nerka .... sockeye salmon
Oncorhynchus tshawytscha .... chinook salmon
Salmo salar .... Atlantic salmon
Salmo trutta .... brown trout
Salvelinus alpinus .... Arctic char
Salvelinus confluentus .... bull trout
Salvelinus fontinalis .... brook trout
Salvelinus namaycush .... lake trout
Thymallus arcticus .... Arctic grayling
Morone americana .... white perch
Morone saxatilis .... striped bass
Lepomis macrochirus .... bluegill
Micropterus dolomieu .... smallmouth bass
Micropterus punctulatus .... spotted bass
Micropterus salmoides .... largemouth bass

---

[a]Names according to: American Fisheries Society. 1991. Common and scientific names of fishes from the United States and Canada, 5th edition. American Fisheries Society, Special Publication 20.

CONVERSION FACTORS

English to metric	Metric to English

Linear measure

English to metric	Metric to English
1 in = 2.54 cm	1 μm = 0.000039 in
1 ft = 30.48 cm; 0.30 m	1 mm = 0.039 in
1 yd = 0.91 m	1 m = 39.37 in; 3.28 ft; 1.09 yd
1 mi = 1.61 km	1 km = 3,281 ft; 0.62 mi

Square measure

English to metric	Metric to English
1 acre = 0.40 ha; 43,560 ft^2	1 m^2 = 10.76 ft^2; 1.20 yd^2
1 mi^2 = 2.59 km^2	1 ha = 2.47 acres
	1 km^2 = 0.39 mi^2

Cubic measure

English to metric	Metric to English
1 ft^3 = 0.03 m^3	1 m^3 = 35.31 ft^3; 1.31 yd^3
1 yd^3 = 0.76 m^3	

Liquid measure

English to metric	Metric to English
1 qt = 0.95 L	1 L = 1.06 qt

Mass

English to metric	Metric to English
1 lb = 0.45 kg	1 mg = 0.000035 oz
1 ton = 2,000 lb = 0.91 metric ton (t)	1 g = 0.035 oz
	1 kg = 2.20 lb
	1 t = 2,200 lb; 1.10 tons

Combinations

English to metric	Metric to English
1 ft^3/s = 0.03 m^3/s	1 m^3/s = 35.31 ft^3/s
	1 g/m^2 = 8.92 lb/acre
1 lb/acre = 1.13 kg/ha	1 kg/ha = 0.89 lb/acre
1 ton/acre = 2.27 t/ha	1 t/km^2 = 2.82 tons/mi^2
	1 mg/L = 0.000016 oz/qt

SYMBOLS
AND ABBREVIATIONS

The following symbols and abbreviations are used in this monograph.

°C	degrees Celsius or centigrade
d	day
ft	foot
g	gram
ha	hectare
in	inch
kg	kilogram
km	kilometer
L	liter
lb	pound
m	meter
mg	milligram
mi	mile
mm	millimeter
μm	micrometer
oz	ounce
ppm	parts per million
qt	quart
s	second
t	metric ton
yd	yard

INTRODUCTION

The presence of sediment is one of the most obvious characteristics of small streams. Sediment has several forms and sources, but of greatest concern in stream and river sediment problems are the fine inorganic particles that either flow with the current (causing turbidity) or that are deposited on the streambed (causing loss of benthic productivity and fish habitat). Such sediment is widespread and pervasive, occurring to some extent in all streams.

The main sources of inorganic sediment are the erosion of uplands, lateral movement of channels into streambanks, and downcutting of streambeds. Evidence of natural erosion is present almost everywhere—hills, valleys, and canyons sculpted by wind and water. Erosion carves down mountains through geological time or, on a shorter time scale, washes out deep gullies in a single storm. Natural erosion results from precipitation in several forms, including secondary events such as floods, melting snow packs, or glaciation. It creates both the spectacular canyons of the arid West and the shallow channels of small brooks.

On its way to the sea, sediment may be temporarily delayed and deposited, only to be eroded and washed out again. Retention times of silt and sand vary greatly. Deposits that we see in streambeds often remain only for short periods of time, perhaps a year or even only a few days. At another extreme, deposits resulting from past glacial melting may be present for so long that to us, they appear as a permanent part of our landscape.

Compared with the huge deposits of geologic history, however, most natural sediment inputs are very small and can be incorporated by stream processes into nondestructive forms and quantities. It is *excessive* sediment, generated as anthropogenic waste, that often overwhelms the "assimilative capacity" of a stream (Cairns 1977) and damages its biological components.

Human influence has been an accelerating factor in modifying the North American environment for only about 300 to 400 years. Obvious effects of such anthropogenic erosion and sediment deposition include loss of agricultural soils, decreased water-retention capacity of forest lands, increased flood frequency, and rapid filling of reservoirs. Less obvious, however (and until recently largely ignored), is sedimentation in small streams that affects biotic communities, reduces diversity of fish and other animal communities, and lowers the productivity of aquatic populations.

A common source of anthropogenic sediment production in the agricultural Midwest—cultivation too close to a stream edge, virtual loss of the riparian zone, and crumbling streambanks. Boot Creek, Minnesota.

OBJECTIVES

The ultimate objective of this review is to encourage more effective management of sediment inputs to streams and to preserve biological integrity and productivity. The chief pragmatic goal is to assist in the improvement and maintenance of stream fisheries, but for other societal interests as well.

Specific objectives are to: (1) identify the main causes or sources of anthropogenic inorganic sediment, (2) summarize the results of recent research on the effect of sediment upon stream biota, and (3) describe sediment control measures aimed at the preservation of viable stream communities and freshwater fisheries.

EXTENT OF SEDIMENT POLLUTION

The amount of sediment produced by human activities is immense. Because of this quantity, sediment is recognized as the major pollutant of United States waters. Oschwald (1972), in a review of a U.S. Department of Agriculture Report to the President (USDA 1969), stated: "Sediment originating from soil erosion has been called the major pollutant of surface waters in terms of quantity involved." Ritchie (1972) began his review with a similar statement: "Quantitatively, sediment is the greatest single pollutant in the nation's water." Norman Berg, U.S. Soil Conservation Service (USSCS), agreed, stating in a U.S. Environmental Protection Agency (USEPA) symposium on urban erosion in 1979 (Downing 1980)

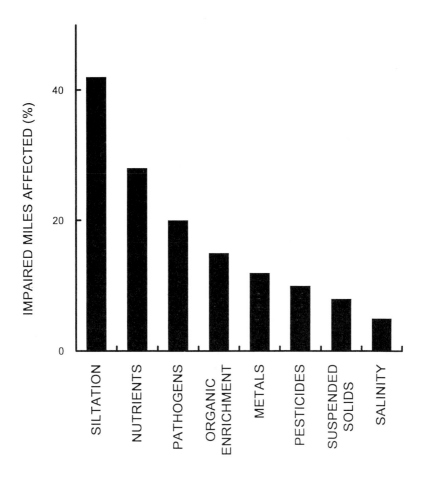

IMPAIRED MILES AFFECTED (%)

FIGURE 1.—Relative importance of pollutants affecting United States streams according to the number of stream miles degraded (from USEPA 1990).

that stream-deposited sediment was the greatest single water pollutant; Berg estimated that two billion tons of sediment were being deposited in U.S. streams annually. Peter C. Myers (1982), Chief of the USSCS, repeated the figure of two billion tons per year. With concern primarily for fine-particle clay erosion on urban construction sites, Lemly (1982) pointed out that " . . . sediment appears to overshadow all other pollutants in both overall quantity added to the receiving waters and total economic and ecological impact."

In a national survey conducted in 1982, the U.S. Fish and Wildlife Service (USFWS) concluded that excessive siltation was the most important factor adversely affecting the nation's fishery habitat in streams (Judy et al. 1984). In a recent USEPA inventory of stream water quality, siltation was identified as the most important cause of river and stream pollution—about 50% greater than the next most important cause in terms of stream distance degraded (USEPA 1990; Figure 1).

Surface mining of metal minerals produces immense spoil piles and tailings subject to erosion. Across abandoned tailings from a stibnite mine (the major source of antimony) in Idaho, a previously diverted small stream now incises and erodes sediment. (Photograph courtesy of William S. Platts.)

Similar reports and conclusions relating to sediment conditions in Canada have not been forthcoming, apparently because few rivers in Canada carry such large quantities of sediment (Stichling 1974). Nevertheless, literature included in this review describes similar severe problems in Canada.

EARLY STUDIES

The earliest published comment on the potentially destructive effects of anthropogenic sediment in streams may have been that by the ichthyologist David Starr Jordan (1889), who remarked on the loss of trout habitat in Colorado streams due to sediment from mining activities. Several other early warnings were forcefully issued in the first few decades of the next century, for example, Titcomb's (1926) indictment of deforestation in New England that caused heavy losses of brook trout habitat; the concern about "erosion silt" expressed by Ellis (1936); and the loss of biological productivity in streams noted by Moore (1937). Despite these early warnings, regulatory agencies did not acknowledge silt and sediment as pollutants of natural waters until the mid-1900s.

In the 1940s, the U.S. departments of Agriculture, the Army, and the Interior brought together an inter-agency committee to coordinate work on water resources. Several subcommittees were formed, including one on sedimentation. Three conferences were held to exchange information on sediment data, sediment sampling methodology, criteria, results of research, and other aspects of sedimentation. Proceedings of the first meeting held in 1946 in Denver, Colorado, were published by the U.S. Bureau of Reclamation (1948); the second conference was held in Jackson, Mississippi, in 1963, and proceedings were published by the U.S. Agricultural Research Service (USDA 1965). Proceedings of the third conference, held in 1976, were prepared by the newly formed U.S. Water Resources Council (USWRC 1976) and were by far the most detailed and comprehensive publications on sediment up to that date. This third conference included participants from 13 federal agencies. The group of interested agencies was officially established as the Inter-Agency Committee on Water Resources (IACWR) in 1955, and the IACWR was superceded by the U.S. Water Resources Council (USWRC) in 1966; its Sedimentation Committee became responsible for later conferences and published proceedings. Together, the three conferences and their proceedings marked a major contribution to the literature on sediment as a pollutant. Several of the proceedings papers are cited in this review.

In 1957, the U.S. Public Health Service began a program in the Pacific Northwest, by which those persons and organizations engaged in water pollution research could be better served through improved communication. Entitled "Technical and Research Consultation Project," the program's major objectives were to "encourage, guide, coordinate and develop research" in the area of water pollution. Among other activities, a series of symposia was held, the fifth of which was "Siltation—its Sources and Effects on Aquatic Life" (Eldridge and Wilson 1959). In addition to a field session on sampling of conditions in salmonid redds and collection of benthic invertebrate samples, discussion topics included turbidity and photosynthesis, siltation and fish-food organisms, siltation and fish spawning, and related subjects. For some time in the sediment-related literature, these proceedings were referred to informally as "the Fifth Symposium." It was also one of the first major publications to include the recognition of anthropogenic sediment as an important stream pollutant. Many more such symposium and conference proceedings were to follow.

BIOTIC STREAM COMMUNITIES

Three major biotic communities are addressed in this volume: (1) photosynthesizing plants, or primary producers; (2) benthic invertebrates, mainly secondary producers in the zoobenthos; and (3) fishes, including both secondary consumers and top carnivores or piscivores.

Photosynthetic plants include the algae, mainly diatoms, that make up the periphyton on submerged stones and other substrates. These primary producers sustain grazing invertebrates such as some of the aquatic insects and mollusks. Other primary producers are aquatic macrophytes such as mosses, watercress, and other rooted plants. In general, these macrophytes are not grazed directly by animals but serve an important function in contributing large quantities of detritus when the plants die back; the detritus then serves as food to detritivorous animals. Potential effects of sediment on aquatic plants include abrasion of periphyton by medium-sized particles such as sand, uprooting of macrophytes by larger particles, smothering of periphyton and rooted plants by heavy fine-

Clear-cutting in mountain forests with steep hillslopes often has been blamed for excessive sediment production. Extensive research, particularly in the Pacific Northwest, has indicted the logging road as the major sediment source. Baker Creek, Idaho. (Photograph courtesy of William S. Platts.)

sediment deposition, and reduction by high turbidity of the light necessary for photosynthesis.

Invertebrates that make up the zoobenthos include many aquatic insects such as immature mayflies (Ephemeroptera), stoneflies (Plecoptera), and caddisflies (Trichoptera), groups often collectively abbreviated as EPT. The abundance of these three orders is commonly combined as an index to availability of food for stream fishes (Lenat 1988). Other major invertebrates include insects such as midges (Chironomidae: Diptera); crane fly larvae (Tipulidae: Diptera); some other dipterans; beetle larvae (Coleoptera); crustaceans including scuds (Amphipoda), sow bugs (Isopoda), and crayfish (Decapoda); worms (Tubificidae: Oligochaeta); Mollusca including snails (Gastropoda) and clams (Bivalvia); and a few other taxa occasionally abundant in certain conditions.

In addition to these macroinvertebrates, some microcrustaceans (Copepoda, Cladocera, Ostracoda) are of increasing concern (Shiozawa 1986). Little ecological information on these forms is available, apparently because they have only recently been recognized as a potentially important part of the benthic animal community in small streams.

Potential effects of sedimentation on benthic macroinvertebrates include interference with respiration and the overwhelming of filtering insects such as some caddisfly larvae that employ fine-meshed catchnets for obtaining drifting food particles. However, the major effect upon benthic invertebrates is the massive smothering of physical habitat by heavy sediment deposition on the streambed, including the loss of interstitial space occupied by burrowing or hyporheic animals.

Effects on fishes have elicited the most concern relating to sediment. Most published work in the literature has been directed toward trout and salmon (Salmonidae), but recently interest has been increasingly directed toward other fishes, particularly warmwater sport fishes and fish species in Endangered or Threatened status. The major threat to fish by sediment is to their reproductive success and loss of rearing habitat. Developing eggs and newly hatched fry may be killed by suspended sediment through interference with respiration or with emergence from spawning gravels; extreme cases of fine sediment concentrations may cause visual impairment in feeding behavior. Deposited sediment fills interstitial space essential for the winter cover of fry and destroys pool habitat used by juvenile and adult fish.

SCOPE OF REVIEW

This review is concerned exclusively with the biological effects of inorganic sediments, introduced as pollutants into streams directly or indirectly through human activities.

Sediments addressed are those that negatively affect living organisms, and range in size through clay, silt, and sand. Finer sizes of particles have generated the most concern and research, especially owing to their more subtle effects. Larger particles, such as gravel and cobble, are usually positive in their biological effects. Larger particles can be negatively important, however, in an aggrading stream reach, where they may decrease channel depth and direct erosive forces laterally to widen the channel and erode banks.

The erosion of streambanks occurs naturally on many streams, but it is often exacerbated by row-crop cultivation and livestock trampling. Its ubiquitous nature and widespread occurrence in almost all streams makes streambank erosion one of the most difficult of all sediment sources to control. Little Cannon River, Minnesota.

Most past and current field research has focused on salmonid streams that support either recreational trout fisheries or juvenile salmon of commercial importance. Because of the usual location of salmonid streams, sediment sources are often linked to forestry practices or mining activities in the steep hillslope topography of western North America and the Appalachian Mountains, or to agricultural practices in some areas of the Great Lakes region and southeastern United States.

Much less information is available from nonsalmonid streams, generally classed as "warmwater." In such streams, smallmouth bass and related centrarchid fishes, catfish, and other warmwater species are of principal interest. Most warmwater streams occur in the midwestern or southeastern U.S. regions, where agricultural practices appear to be the main source of high sediment inputs (Striffler 1964; Lenat 1984; Menzel et al. 1984; USEPA 1990). Concern for these streams and research on their biological communities are increasing.

This review focuses on information from studies in North American streams and rivers. However, relevant information from other regions, particularly when concerned with principles (as opposed to local results) is also included. Some general problems in streams that may be considered ecological counterparts in their biology are apparent in Australia and New Zealand, western Europe, South Africa, and other locations. Substantive results from ecological research on stream sedimentation have appeared in the published literature only in the last two decades.

The Appalachian Mountains are a rugged region of steep hillslopes and high stream gradients, exemplified by this healthy watershed of Bear Wallow Creek, North Carolina. The steep topography, however, holds great potential for severe sediment erosion when human activities, such as logging, cultivation, or recreational home development, disturb land surfaces.

Floodplain cultivation has been labeled one of the major anthropogenic sources of sediment, especially in warmwater streams. Fall plowing that leaves tilled soil exposed to spring floods, with straight furrows leading directly to stream waters (behind trees in background) provides an important sediment source should floodwaters inundate these fields. Zumbro River, Minnesota.

Considerable interest in erosion and the consequent effect of soil losses upon agricultural production was apparent earlier than were ecological concerns. Guy and Ferguson (1970) pointed out that we are in the second period of concern about the effects of soil erosion. The first of these periods (beginning in the 1930s), which included great flood-generated losses of soil, also resulted in the creation of government agencies concerned with the loss of agricultural productivity (e.g., the U.S. Soil Conservation Service, U.S. Agricultural Stabilization and Conservation Service). The second period of concern includes the problem of detrimental effects of the sediment (or "soil losses") upon biological communities in receiving waters. The agriculture industry emphasizes the loss of productivity from erosion, rather than the pollutional character of the sediments transported to streams; for example, the U.S. Soil Conservation Service establishes a certain level of tolerance of soil losses (called T values; e.g., 5 tons/acre per year) that is acceptable from the agricultural viewpoint (Cook 1982). However, the technologies developed by agriculture for preventing soil losses can just as well be employed for preventing the biologically damaging effects of sediment transported to streams.

Unrestricted access to streams by cattle results in the destruction of the buffering capacity of the riparian zone, causing slumping streambanks, widened and shallowed channels, loss of soil productivity through erosion—and sedimentation of the streambed. Small stream in southeastern Minnesota.

PREVIOUS PUBLICATIONS ON SEDIMENT

The U.S. Department of Agriculture's Yearbook of Agriculture, *Water* (USDA 1955), included an early summary of general effects of erosion and sediment transport; it apparently was a major stimulant to future research. Expressions of concern also appeared in popular publications about the same time, chiefly involving stream fisheries (Wolf 1950; Eschmeyer 1954; Kelley 1962).

The first major literature review on the effects of anthropogenic stream sediments was by Cordone and Kelley (1961): "The Influence of Inorganic Sediment on the Aquatic Life of Streams." This seminal publication stimulated broader environmental concerns for the ecological integrity of stream and river communities, combined with increased awareness of damaged fisheries. Rigorous field research and critical experimentation followed in subsequent literature, generally beginning in the 1970s. In his definitive reference volume on stream ecology, Hynes (1970) included a chapter on effects of human activity on streams, including soil erosion and silting of streams. Hynes (1973) later published a brief general review on sediment that affects stream biota.

Most review articles published in recent years have been informal publications on specific subtopics, or about limited geographic areas. Although many of these articles have been used in preparation of this review, especially for literature references, and are themselves cited in appropriate locations, the analysis and interpretation in this review are based almost entirely on primary material.

Normally dry headwater flowages, maintained with healthy vegetative cover, contribute greatly to the control of erosion during snow melt and summer storms. Such cover assists in prevention of gullies and sediment transport to lower streams and rivers.

Major reviews include those by Wallen (1951), Cordone and Kelley (1961), EIFAC (1965), Cummins (1966), Chutter (1969), Bjornn et al. (1974), Sorensen et al. (1977), Iwamoto et al. (1978), Muncy et al. (1979), Hall et al. (1984), Reiser et al. (1985), Garman and Moring (1987), Lloyd (1987), Salo and Cundy (1987), Seehorn (1987), Chapman (1988), and Platts (1991). Full reference data for all reviews are given in Appendix I.

PHYSICAL CHARACTERISTICS OF STREAM SEDIMENTS

The size of sediment particles is an extremely important attribute in the effect produced upon stream communities. A grade scale developed by the American Geophysical Union comprises 24 categories of sizes and class names, ranging from "very fine clay," 0.24 to 0.5 μm, up to "very large boulders," 2,048 to 4,096 mm (80 to 160 in) (Lane 1947). However, a classification of sediment categories more commonly used by biologists and stream ecologists was provided by Cummins (1962), with 11 particle sizes and names (Table 1), based on the Wentworth scale (Wentworth 1922). The size classes range from clay (the smallest) upward through silt, sand (five classes), gravel, pebbles, and cobbles to boulders (the largest).

Fluvial geomorphologists, in considering sediment transport in streams and rivers, divide the solid sediment load into the categories of *suspended load*, and *bed load* (Richards 1982). Suspended load consists of small particles—but sometimes ranging up to coarse sand—that are carried along in the water. Although suspended sediment is the product of eroded sediment inputs from outside the channel, its maximum concentration is limited by water velocity and turbulence. Most of the suspended load is *wash load*, defined as the finest particle sizes that originate in the watershed and which are not normally found in streambed deposits. Bed load consists of larger particles on the stream bottom that move by sliding, rolling, or saltating along the substrate surface. Particles of intermediate sizes (usually fine sand) may interchange between the two defined components depending on water velocity and other water characteristics.

These definitions do not necessarily mean that fine particles will not settle in turbulent water. For example, clay-sized particles may deposit on epilithic periphyton on stones, particularly on algal cells that may have sticky, mucilaginous surfaces (Graham 1990).

Biologists recognize the two components of sediment load by their respective effects upon stream organisms, both plant and animal. Thus, this volume is partially organized by suspended sediment and deposited sediment, because these two components respectively affect the flowing water and the streambed substrate, the two major parts of the environment of stream organisms. Rather than the rate of transport, biologists will be mainly interested in (1) the *concentration* of suspended sediment (related to suspended load transport), and (2) the degree of *sedimentation* on the streambed (related to bed-load transport).

TABLE 1.—Sizes of sediment particles (modified from Cummins 1962).

Category	Size range	Phi scale
Boulder	>256 mm	−8
Cobble	64–256 mm	−6, −7
Pebble	16–64 mm	−4, −5
Gravel	2–16 mm	−1, −2, −3
Very coarse sand	1–2 mm	0
Coarse sand	0.5–1 mm	1
Medium sand	0.25–0.5 mm	2
Fine sand	0.125–0.25 mm	3
Very fine sand	0.0625–0.125 mm	4
Silt	4–62 μm	5, 6, 7, 8
Clay	< 4 μm	9

The concentration of suspended sediment affects the light available to photosynthesizing plants and the visual capability of aquatic animals. Streambed sedimentation affects the habitat available for macroinvertebrates, quality of gravel for fish spawning, and amount of habitat for fish rearing. Suspended sediment may be measured in terms of concentration of total particulate solids in parts per million (ppm, = mg/L). The degree of streambed sedimentation is more difficult to measure, especially in terms that would have biological meaning. Some possibilities include (1) area of streambed covered, (2) depth of coverage, (3) percent of defined "fines," and (4) percent saturation of interstitial space or embeddedness.

Usually, the size separating suspended sediment from deposited sediment is about 62 μm; that is, suspended sediment comprises mostly clay and silt, but under conditions of greater slope and water velocity, fine-to-medium sand may also be entrained in suspended sediment. Deposited sediment may be composed of clay, silt, some grades of sand, and gravel. Biologists do not consider the largest particles, such as cobble and boulders, as pollutants, because these two sediments are generally beneficial to stream organisms.

The two categories, suspended sediment and deposited sediment, should not be considered mutually exclusive as to particle size, because they will overlap considerably depending on conditions of current velocity and water turbulence. The convention is useful, but it should not be construed to suggest a static condition. Both suspended sediment and deposited sediment continue in motion, slow or rapid, to cause stream meandering, streambank erosion, and bar formation, as well as creation and loss of fish habitat.

Inorganic particles down through the size of silt are derived largely from the physical breakdown and weathering of silicate rocks. Clay particles, however, are synthesized chemically in layers formed with crystals of aluminum and magnesium silicate (Hassett and Banwart 1992). Whereas clay crystals occur in sizes less than 2 μm, aggregations of crystals form larger particles. In natural streams, predominant inorganic particles range from 10 to 20 μm, being aggregates of clay and silt (Walling and Kane 1984).

In contrast to larger sizes, some clay particles have a very high ion-exchange capacity and thus may carry large quantities of adsorbed nutrients and contaminants (subjects not covered in this review). Clay particles, small and somewhat adhesive, are extremely pervasive and may be responsible for many of the destructive effects upon stream biota. Research on the biological effects of clay specifically, however, has not been reported extensively.

The U.S. Soil Conservation Service has published engineering data relating to sediment, including definitions, physical properties, water transport, yield, delivery, and other basic aspects of quantitative expression for sediment generation and behavior (USSCS 1983).

SOURCES
OF SEDIMENT

Natural stream erosion, in the absence of human activities, is affected by various parameters of water flow and channel morphology, in combination with the type of catchment bedrock, soil profiles, and vegetation (Leopold et al. 1964). Anthropogenic sources, on the other hand, include agriculture, forestry, mining, and urban development, among others (USEPA 1990).

More eclectic in distribution, however, are streambank erosion and service roads—extremely important as sources of sediment in almost all land-use operations. Although a natural process, streambank erosion often erodes floodplains composed of previously deposited sediment of anthropogenic origin. Streambank erosion also can be accelerated by human activities, including agricultural encroachment, timber harvest in the riparian zone, and heavy recreational use. Akin to streambank erosion is erosion of streambeds caused by incision; streambed erosion also can be increased by anthropogenic activities that alter stream morphology, such as damming, channelization, and gully formation due to agricultural tillage.

Roads that service resource extraction or development operations, especially in the construction stage, have been tagged as the most acutely responsible for eroded sediment. Miscellaneous activities, such as bridge and dam construction, pipeline and electric transmission crossings, and virtually any other activity that involves disturbance of the land surface can generate sediment, which may then be transported to streams.

The main purpose of this chapter is to identify and describe the various sources of anthropogenic sediments that are likely to enter and degrade small streams. Many literature references that include source information, but which emphasize specific effects, are included in later sections.

AGRICULTURE

Among all sources of pollution afflicting streams and rivers, agriculture in its several forms is by far the most important—over three times the amount of pollution contributed by the next leading source (USEPA 1990). Published reports on the effects of agricultural practices that produce sediments cite row-crop cultivation on floodplains and livestock grazing in riparian zones as principal sources. Erosion from cultivation is more important in midwestern and southeastern regions (affecting mainly warmwater stream fisheries); livestock grazing is more important in western regions (affecting mainly salmonid fisheries).

Row-crop Cultivation

The production and transport of sediment to streams, among all sources, is greatest from row crops and other cultivated fields (Costa 1975; Lenat et al. 1979; Clark 1987). Robinson (1971), director of the U.S. Department of Agriculture Sedimentation Laboratory, Oxford, Mississippi, concluded that agriculture contributed about one-half of all sediment pollution in the United States.

Fajen and Layzer (1993) drew a sharp contrast between presettlement conditions in the Midwest and Southeast United States, and conditions following agricultural development by European immigrants. In prehistory these regions flourished with productive soils held in place by native vegetation and a water transport system that stored much water on the land and broadly distributed natural sediments. Upon settlement and agricultural development, friable topsoils were exposed to summer rainstorms characteristic of these regions, causing extreme erosion, head-cutting gullies, and streams laden with sediment.

A particular indicator of agricultural erosion and its resultant generation of sediment transported by streams is the accretion of floodplains that has buried the original soil profiles. Since European settlement and commencement of agricultural cultivation in the early to middle nineteenth century, such alluvial accretions occurred in the Upper Mississippi Valley on the order of 0.3 to 4.0 m (Costa 1975; Trimble and Knox 1984).

Some of the earliest literature that dealt with losses of stream fish and invertebrate populations concerned sediment eroded from soils in midwestern agricultural regions (Trautman 1933; Aitkin 1936; Ellis 1936; Larimore and Smith 1963). These reports, however, were broad observations that mainly associated changes in fish species and abundances with visually obvious muddy water or, alternatively, with the appreciation of known large soil losses from croplands. There has not been much detailed sampling or experimental work conducted on specific biological effects of siltation from cultivated fields on invertebrate production, warmwater fish reproductive success, or other biotic responses.

Later warnings and appeals appeared commonly in the popular media, continuing to the present, about the damaging effect of sediments on recreational fisheries in the agricultural Midwest (Eschmeyer 1954; Reinert and Oemichen 1976; Matthews 1984). Sedimentation resulting from agricultural sources has been indicted as the major environmental problem in warmwater streams (Judy 1984; Roseboom et al. 1990). Agriculture has been charged as the activity most responsible for loss of fish species in midwestern streams, particularly by causing turbidity (Karr et al. 1985).

Agricultural science, however, has made great advances in erosion control on row-crop fields. The literature on agriculture teems with reference books, symposia, guidebooks, and manuals aimed at the measurement and control of soil losses from cropland erosion (Lum 1977; Batie 1983; Overcash and Davidson 1983; Beasley et al. 1984; Harlin and Berardi 1987; Wittmuss 1987; Young and Onstad

Straight furrows, up-and-down slopes, in rolling terrain results in inevitable loss of soil and eroding sediment. Small stream flows to right in back of camera. Southeastern Minnesota.

1987; Schwab et al. 1993). Many technologies have been designed to greatly reduce runoff: contour plowing and planting, conservation tillage, strip cropping, crop rotation, terracing, leaving crop residues, and others. The extent of implementation and the resulting success are less known. Furthermore, some erosion appears inevitable; tolerable levels (T values), expressed by the U.S. Soil Conservation Service as tons per acre, are considered losses that can be sustained indefinitely (by continuing soil development) without reduction of soil productivity (Lum 1977; Cook 1982; Batie 1983).

Some attention has been given to the economic assessment of off-site environmental damages, including reduction of stream-based recreation (Clark et al. 1985; American Farmland Trust 1986; Ribaudo 1986; Waddell 1986; Clark 1987). None of this literature, however, addresses specific damage to the stream biota upon which most stream recreation (sport fishing) depends. Even less has been investigated and published by water pollution biologists and resource scientists on effects of sediment on biological communities in warmwater streams.

The most common approach to identify and evaluate sources of sediment from row-crop erosion has been to compare different streams or stream reaches variously affected by cropland runoff. Selected examples, from both United States and Canadian agricultural areas, exemplify these comparisons. Yet, whereas reductions in primary producers, macroinvertebrates, and fish have been recorded in those stream communities most affected by agricultural runoff, the specific causal factors are missing. The question—How exactly do sediments affect these biological elements?—has not been extensively addressed.

In the United States, Peters (1967) examined the fish population in Bluewater Creek, Montana, a trout stream that flowed through an area intensively developed for cropland and pasture. He noted that as sediment levels increased fivefold through the stream's course, brown trout decreased from more than 2,000 fish/acre to less than 100/acre and that numbers of nonsalmonid fishes (minnows, suckers) increased by more than 100 times.

In North Carolina, several sets of streams were compared for effects on invertebrates. The streams were on lands subjected to different levels of agricultural use. When compared with other streams in undisturbed or forested watersheds, the orders Ephemeroptera, Plecoptera, and Trichoptera (EPT) were lacking in streams affected by agriculture, and the more tolerant Chironomidae were more abundant (Lenat 1984; Crawford and Lenat 1989).

In the same region, Braatz (1993b) demonstrated highly significant increases in suspended sediment below a farming operation. Although average baseflow suspended sediment concentrations were not changed, storm flows were largely responsible for overall increases. Average stormflow suspended sediment rose from 185 mg/L upstream from the farm to 1,640 mg/L downstream, about an 8.5-fold increase. These results were attributed to degraded streambanks trampled by livestock and lack of buffers between tilled fields and raw, exposed slopes leading to the stream.

In southern Minnesota streams, Troelstrup and Perry (1989) found fewer macrophytes and lower macroinvertebrate densities in reaches associated with cultivated riparian areas than reaches within forested areas. They also reported smaller substrate particle sizes, higher gross primary production (benthic), and higher levels of pesticides in the agriculture reaches.

Wilkin and Hebel (1982) studied the distribution of eroded sediment in the agricultural watershed of the Middlefork River, Illinois, and concluded that sediment delivery to streams was increased by row-crop cultivation on floodplains.

Welch et al. (1977) compared fish and invertebrates between two groups of New Brunswick streams—one group in croplands, one control—and reported that numbers of trout and sculpins were greatly reduced in the agriculturally impacted streams, mayflies and stoneflies were reduced, and chironomids had increased.

Dance and Hynes (1980) compared the benthos between two main branches of Canagagigue Creek, Ontario, one branch in an agricultural watershed, and one branch in a forested area. They reported lower species richness, fewer of the shredder functional group of insects, and more chironomids in the agricultural stream branch.

In yet another study of aquatic invertebrates, Cobb and Flannagan (1990) studied emergent caddisflies from the Ochre River, Manitoba. Sedimentation resulting from land clearing, channelization, and stream straightening reduced both species diversity and abundance.

No published account was found on stream sediment damages from croplands in the plains provinces of Canada, but to the extent that row crops are cultivated in that region, the potential for damage from agricultural sources probably would be similar.

Ironically, it is the region of degraded warmwater fisheries that has received the least scientific inquiry into the biological effects of sediment. For example, where overall damage to stream biota has been investigated, specific factors responsible are usually unknown.

Two major difficulties arise when using the above broad comparative approach to establish either the identity of a sediment source or its specific effect. The first difficulty is that sediment input from cultivated fields to streams is a form of nonpoint pollution and therefore extremely difficult to measure and evaluate (Lenat et al. 1979; Simons and Li 1983). With nonpoint sources, stream degradation from sediment generally can not be distinguished from other effects of agricultural practice, such as elevated temperatures and toxic pesticides. Even when the effect of sediment by itself may be isolated, further inquiry should address the mode of action, for example, size-particle composition of the substrate and embeddedness, and their effects on fish reproductive success and invertebrate production. Among all sources of inorganic sediments as pollutants, the agricultural source now requires the greatest attention in both specific descriptive and experimental research on biological responses.

The second difficulty is that many midwestern streams already have been changed from their pristine condition so that it is rare to find reaches unpolluted by sediment and other runoff elements. A striking example is the study of known sedimentation that took place nearly a century ago in the Kankakee River, reported by Brigham et al. (1981). The Kankakee River originates in an upstream reach in Indiana, where it was channelized and drained for agricultural use in the early 1900s. The downstream reaches in Illinois, laden with heavy deposits of sand and silt, were thought to have undergone losses in aquatic life as the result of sedimentation. Indeed, the authors' findings included fewer species and lower abundances of fish, mussels, and benthic invertebrates in the areas with higher sand–silt substrates than in the fewer available reaches less affected. However, as in other studies, it was impossible to identify specific sources or how sediment produces specific effects on the biota.

The research of Berkman and Rabeni (1987) appears to be the only study addressing the effect of sediment on specific functional aspects of biological response to cropland-derived sediment. Commenting on the previously cited reports of subjective conclusions relating fish species changes to agriculture, Berkman and Rabeni cited the need " . . . to understand the biological basis for the effects." Based on three streams in northeast Missouri in which they sampled substrate composition and fish distribution, they concluded that the major effect of siltation was the reduction in density within simple, lithophilous reproductive guilds—fishes that require clean gravel or cobble substrate for spawning. These fishes included the golden redhorse, white sucker, and several cyprinids.

By all accounts, the greatest source of stream pollution by sediment in warmwater streams is soil erosion from cultivated croplands. Although specific sediment sources and functional biological effects have not been adequately identified, it remains clear that past damage to stream fisheries, as recorded in subjective observations long ago, has been extensive in the numerous warmwater streams of the Midwest.

Livestock Grazing

Overgrazing by livestock on arid rangelands has been responsible for immense damage to streams in the western United States. Such damage is due largely to the generation of excess sediment caused specifically by livestock overuse of stream riparian areas.

At great risk are salmonid spawning reaches used by anadromous Pacific salmons and inland trouts. Unrestricted use by livestock, mainly cattle, of the richer, moister forage in streamside areas results in a number of degrading effects. Principal among these is the trampling of streambanks and the consequent loss of streambank stability. In addition to sedimentation, degrading effects include elevated stream water temperatures and loss of undercut banks used for cover and fish-rearing habitat.

The overgrazing problem in western rangelands, and its effects on salmonid fisheries, have been recognized for several decades. One of the earliest papers to express concern about the effects of grazing on a stream's fishery (although the factor of sediment was implicit) was that by Tarzwell (1938), who reported fencing-enclosure experiments in the southwestern United States.

In later decades, however, many conferences, symposia, perspectives, forums, review articles, and other published summarizations of the problem were added to a burgeoning research literature on overgrazing (Johnson and Jones 1977; USFWS 1977; Meehan and Platts 1978; Cope 1979; Johnson and McCormick 1979; Moore et al. 1979; Platts 1979, 1981b, 1982; Menke 1983; Kauffman and Krueger 1984; Johnson et al. 1985; Armour et al. 1991). Exemplary individual papers include those by Meehan et al. (1977), Behnke and Raleigh (1979), Platts and Martin (1980), Platts (1981a), Hubert et al. (1985), Rinne (1985), and Stuber (1985).

In an insightful review related to fisheries, Platts (1981b) summarized the major effects of western livestock grazing on aquatic habitats, including several channel parameters, water quality components, and changes in fish populations. Platts concluded: "Stream-channel sedimentation caused by soil erosion on millions of acres of rangeland has been recognized as a major problem." In a subsequent review of overgrazing effects on stream fisheries, Platts (1982) judged from all reports considered together that "...past livestock grazing has degraded riparian-stream habitats and decreased fish populations." In a more recent major review, Platts (1991) traced the history of grazing on western rangelands, grazing strategies, stream quality factors, and effects on fish populations—a comprehensive summary of the overall problem of western livestock grazing and stream fisheries.

Scientific and management concerns for the riparian zone in arid rangelands continue to increase, as does appreciation of the diversity and productivity of riparian areas. This view was expressed forcefully in the seventh symposium on western wildland shrubs, which was devoted to riparian shrub communities (Clary et al. 1992).

Most effects of overuse of riparian areas include increased fine-sediment generation from channel widening, slowed water and shallowed channels, sediment entrainment from slumping streambanks, and increased deposition on the streambed. Where riparian areas have been denuded by overgrazing, the dam-

Severe overgrazing along waterways and small streams results in slumping banks and water carrying sediment to larger receiving streams.

aging effect of storms (e.g., higher velocities and floods) is much greater in its effect on streambanks and channels (Platts et al. 1985).

In a summary paper on threatened and endangered fish species, Behnke and Zarn (1976) pointed out that good habitat was the best guarantee of sustained native trout populations, but they also emphasized that the influence of livestock grazing on headwater habitat could be devastating to native and threatened species. Destabilization of streambanks by trampling, with the consequent siltation of streambeds, was cited as a major factor. They emphasized the need for more precise data on the effects of livestock grazing on aquatic habitats.

The main experimental approach to identify sources of stream degradation on rangelands has been to compare open streams to streams that have been fenced to exclude livestock (Keller et al. 1979; Van Velson 1979; Claire and Storch 1983; Duff 1983; Marcuson 1983; Stuber 1985). Where a fishery was an important concern, fencing experiments compared fish populations along with various stream parameters. The results from all investigations agreed that significant beneficial effects occurred from livestock exclusion. Those effects were attributed to many factors, including sediment reduction.

However striking the broad comparisons have been, vital questions remain about more specific responses. Although the channel parameters listed above— streambank destabilization and sedimentation—have been often observed, specific effects of these changes have not been intensively evaluated. Platts (1982) and Rinne (1985) also expressed the need for improvement in research design, such as better controls and replications, and greater investment in long-term research. Platts (1981b) recommended that scientists conduct interdisciplinary research to develop new grazing strategies in order to maximize understanding of the aquatic resource among rangeland managers. Probably new grazing strategies would affect all stream channel parameters, promoting higher water quality and fisheries productivity. Improved knowledge of the specific effects of sedimentation would greatly improve the application of management strategies.

It is clear that grazing policies that permit overuse of riparian zones have damaged stream salmonid fisheries in the past. Increased public concern, new governmental appreciation, and administrative agency activities are providing impetus for major reforms of grazing policy, particularly on the huge areas of public land used for private livestock grazing. Research aimed at improved strategies that will permit acceptable livestock production, and at the same time preserve the traditional character and natural values of western salmonid fisheries, is critically needed.

American Fisheries Society scientists made a strong case for protection of riparian areas that affect fisheries in arid-region streams (Haugen 1985). The author urged the implementation of many steps, including a plea to "Insist that riparian area management prescriptions are adhered to by State, Federal, and Provincial land management agencies and that they be monitored for effectiveness." In a later review, Armour et al. (1991) made another strong indictment of livestock overgrazing of riparian systems on western public lands, quoting all of the damaging effects on stream fisheries mentioned above, including sedimentation, and denouncing the American grazing fee system on public lands. They issued an appeal for advocacy by the American Fisheries Society, listing 13 needed points that included public education, research, policy changes, and other reforms.

FORESTRY

Sediment generation from various forestry practices has been investigated extensively. These operations include clear-cutting, skidding, yarding, site preparation for replanting, and road construction and maintenance. The relative contribution of sediment appears to be moderate from clear-cutting (i.e., higher than from selective cutting or patch-cutting), moderately high from skid trails, minimal from yarding (higher if heavy machinery is used near streams), and moderate from site-preparation. By far, excess sediment generation is greatest from logging roads, particularly if built near streams, and much greater if road construction creates conditions for mass soil failures and landslides.

The effect of forest practices on aquatic resources has drawn the greatest attention in land use–fishery resource relationships, particularly in sediment-related problems. Most research has been conducted on streams in steep terrain, notably the mountainous areas of northwestern United States and western Canada, and the southern Appalachian region, because conditions that affect erosion of disturbed land surfaces are greatly intensified on steep hillslopes.

Early Studies and the Identification of Sediment Sources

A few early reports drew attention to some damaging effects of timber harvest on fish (e.g., Titcomb 1926), but it was not until midcentury that concerted research efforts, including critical experimentation, were begun to determine

A road too close. Construction of service roads at streamside (or worse, in the channel itself) can be a major source of sediment in streams, resulting from the need for roads in logging, mining, and other operations in steep topography. Webb Creek, North Carolina.

specific forestry practices that resulted in the observed deleterious effects on stream fisheries and their supporting biological communities (Chapman 1962).

Concern about the influence of sediment and other effects on stream fishes increased greatly in forested regions of the Pacific Northwest and the southern Appalachians, regions with immense standing timber resources and also major salmonid fisheries. These factors set the stage for a profusion of sediment problems, as well as research efforts aimed at the identification of sediment sources. Seminal papers included those by Tebo (1955) in North Carolina, McNeil and Ahnell (1964) in Alaska, Hall and Lantz (1969) in Oregon, Burns (1970) in northern California, and Platts (1970) in Idaho.

In the following decades, investigations into forest management problems documented severe losses in fish populations and fish harvests, and reductions in underlying basic productivity. The relevant literature expanded greatly with experimental studies, development of sampling methods, and tests of control measures. The culmination was the comprehensive reference volume on influences of forest and rangeland management on salmonid fishes and their habitats, edited by Meehan (1991a).

Throughout this quarter century, the recognition of problems and research results expanded in many forest regions—the southern Appalachians, New England and the Canadian Maritime Provinces, and inland streams of the Rocky Mountains, as well as the Pacific Northwest. Because their freshwater habitats are mostly in cool, forested lands at higher elevations, almost all research and publications concerned salmonid fishes. Very little information exists on specific effects of sediment on warmwater fishes in wooded regions, and consequently virtually no forestry-practice sources of excess sediment affecting warmwater streams have been identified.

Many reviews, bibliographies, and summarizations are now contained in the literature on effects of forestry practices including sedimentation (USDA 1965; Sopper and Lull 1967; Krygier and Hall 1971; Rice et al. 1972; Gibbons and Salo 1973; USFS 1975; Anderson et al. 1976; Meehan et al. 1984; Gibbons et al. 1987; Salo and Cundy 1987; Seehorn 1987; Seyedbagheri et al. 1987; Macdonald et al. 1988; Campbell and Doeg 1989; and the latest by Meehan 1991a). Some reviews emphasized sedimentation from several sources, including forestry practices (Cordone and Kelley 1961; WWRC 1981). Many individual papers in primary journals, agency reports, and informal media have been published on various aspects of forestry practices as sources of sediment affecting stream fisheries.

Long-term Research Programs

Several long-term research programs have been established for many years, in both Canada and the United States. Some research goals have been broad, covering many aspects of forestry practices and many kinds of effects, including effects on aquatic resources; some have been closely focused on aquatic resources, usually salmonid stream fisheries. These long-term projects have been developed mainly in Canada by the federal Department of Fisheries and Oceans, and in the United States by the U.S. Forest Service (USFS).

Some of the earliest work on the effects of logging was done at the USFS Coweeta Experimental Forest (now Coweeta Hydrologic Laboratory), in North

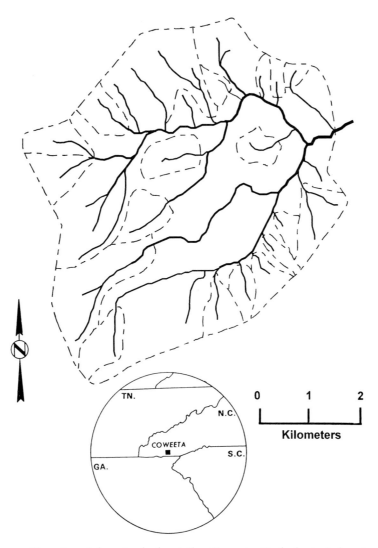

FIGURE 2.—Experimental watersheds at the Coweeta Hydrologic Laboratory, North Carolina (from Swank and Crossley 1988).

Carolina (Swank and Crossley 1988) (Figure 2). This area is divided into small watersheds which are individually treated in experimental forest management programs. The stream in each watershed is regularly monitored for hydrologic parameters, and many are studied for ecological effects of the different experimental treatments.

Among the earliest efforts at Coweeta to define sediment sources and the effects on stream communities was the work of Tebo (1955, 1957), who identified two major sources of sediment in the steep terrain of these southern mountains: logging roads and skid trails. As a consequence, Tebo recommended that roads be built well away from streams and skid trails be operated upslope, rather than down, with the use of cable systems to remove logs. Later work at Coweeta identified sediment input from logging roads in three phases relevant to road crossings over streams: initial construction, periods of major use, and during storms (Webster et al. 1983). Gurtz et al. (1980) further studied the production of sediment by clear-cutting at Coweeta and also attributed it mainly to roads.

Other substantial programs in eastern forests have made important contributions to the search for sediment sources, as well as methods for control. Experimental forestry practices were carried out at the Fernow Experimental Forest, West Virginia, another USFS study area (Eschner and Larmoyeux 1963; Reinhart et al. 1963). Four different logging practices were tested for increased levels of turbidity; the logging methods ranged from commercial clear-cutting through three levels of selective cutting. Resulting turbidity was extremely high with clear-cutting, down to less than 1.0% at the lowest level of selective cutting, and less than 0.1% in a control tract (no cutting). In much of this early research on methods of logging, most concern was expressed about the type of cutting, whereas later work was concentrated on the kind and density of logging roads.

In New Hampshire, the Hubbard Brook Ecosystem Study included research on particulate matter output from the forest before and after deforestation (Bormann et al. 1974). Annual output of particulate matter (about 60% inorganic) increased from an average of 2.5 t/km^2 to a maximum of 38 t/km^2 3 years after deforestation, an increase of about 15-fold.

Other studies in eastern forests confirmed many sources of sediment in forestry practices. Findings from an evaluation of clear-cut logging of a spruce-fir forest on the Piscataquis River watershed, Maine, included increased suspended sediment and high levels of fines in pools, along with other effects—higher temperature and elimination of native brook trout (Moring et al. 1985; Garman and Moring 1991). Studies in the Chattahoochee National Forest, Georgia, confirmed that historic road construction accompanying timber harvest along trout streams had caused major stream sediment problems in the past; these studies resulted in better road design and retention of stream-side buffer strips (England 1987).

In the relatively flat terrain of North Carolina's Piedmont, replanting is practicable after clear-cut harvesting, but site preparation, especially if done mechanically with heavy equipment, can produce heavy sedimentation in streams (Douglass and Goodwin 1980). Several site-preparation techniques were examined for erosion and sediment production: (1) shearing of brush and burning; (2) shearing, burning, and disking; (3) all of the above plus grass seeding; and (4) control. Soil losses were (1) 3,000 lb/acre, (2) 8,000 lb/acre, (3) 640 lb/acre, respectively, but only 10 lb/acre from (4) the control. The authors concluded that mechanical site preparation, which exposed soil and caused visible erosion, was one of the most severe treatments applied to forest land. Douglass and Swift (1977), in an earlier and similar report, had reached the same conclusion.

Most studies to identify sediment sources from timber harvest and other forestry practices have been conducted in the steep hillslope regions of the United States Pacific Northwest and western Canada. Among the most extensive programs of experimental research is the Alsea Watershed Study, located on a tributary of the Alsea River in western Oregon (Figure 3). Hall and Lantz (1969)

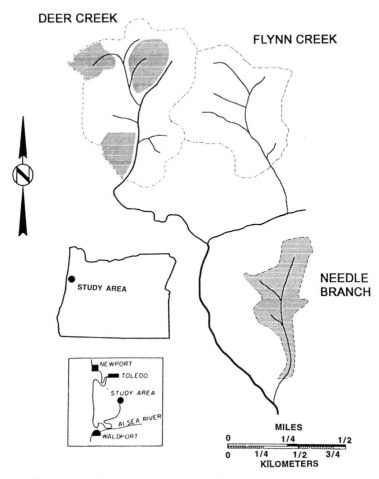

FIGURE 3.—Experimental logging in the Alsea Watershed Study, Oregon: Deer Creek, patch-cut; Flynn Creek, control and uncut; Needle Branch, clear-cut. Shaded areas are experimental cuts (from Moring and Lantz 1975).

first reported biological results from this long-term program. The study comprised three experimental watersheds with different timber-harvest treatments: Needle Branch, clear-cut; Deer Creek, patch-cut with buffer strips; and Flynn Creek, control and uncut. The Alsea study was begun in 1962 and logging was conducted in 1966 on all three watersheds (Brown and Krygier 1971; Krygier et al. 1971). Increased sediment loads in both Needle Branch and Deer Creek were attributed mainly to road construction and (in Needle Branch) also to slash burning that exposed mineral soils, rather than to clear-cutting itself. Later results from the Alsea study emphasized that mass soil movements in Deer Creek (landslides and debris avalanches) associated with road construction were more important than erosion from roads directly. These movements occurred long after logging had ceased and after stable conditions presumably had returned (Beschta 1978). Additional studies in the Alsea area included those by Moring (1982), who reported decreases in stream-gravel permeability of 78% immediately after clear-cutting (remaining depressed by 50% for 6 years). Ringler and Hall (1988) studied the vertical distribution of sediment in salmon redds and reported more fines in the upper layers of both the clear-cut and patch-cut watersheds than in the control.

In the H. J. Andrews Experimental Forest, another forest research program in Oregon, studies also dealt with different timber harvest techniques (Figure 4). Three small watersheds were variously subjected to clear-cutting, patch-cutting, and no logging (control); sediment yield was measured from each. Fredriksen (1970) reported that the clear-cutting with skyline yarding (no roads) produced little sediment; the patch-cut harvest, with high-lead yarding but with many roads, yielded much higher levels of both suspended and bedload sediment. Landslides associated with roads in the patch-cut operation were particularly important sources of sediment.

In northern California, some early concerns about excess sediment from logging and road construction were addressed in the long-term studies initiated in 1962 on South Fork Caspar Creek. Fourteen years of study, during some of which road construction and logging commenced, were reported by Rice et al. (1979); they estimated that sediment yield in the first 5 years after the commencement of logging increased threefold over that which would have been produced for the area if it had instead remained undisturbed.

Additional experimental studies were carried out on four streams in northern California with different logging practices (Burns 1972). Again, roads were strongly indicted in sediment production. In the most severe case, the principal road ran near and along the stream with several bridge crossings, bulldozer yarding was done in the streambed, and road-building materials were side-cast directly into the stream. Deposited sediment increased up to 0.6 m thick, and high concentrations of suspended sediment resulted. In other streams of Burns's study—with high-lead yarding and roads built farther away from streams—much less siltation resulted.

On the basis of data from many private timber-harvest areas in northwestern California, Rice and Lewis (1991) concluded that, by far, most erosion occurs on very small proportions of harvested sites, mainly on roads and in mass soil movements. Their analysis included the development of means to identify those critical sites and improve timber-harvest planning accordingly.

Other major studies on production of sediment and its effects on salmonid fisheries were those from several salmon streams in Alaska that were subjected to sediment pollution from logging operations (McNeil and Ahnell 1964; Sheridan and McNeil 1968). These papers presented some of the first alarms over the severe effects that logging could have on salmonid redds. With this potential hazard to salmon production by sediment realized, greater interest was directed to timber harvest operations in Alaska.

The erodibility of central Idaho's granitic batholith has elicited much study of sediment sources and the severe sediment pollution problems in streams of that region. Long-term research over many years has identified sources and practices generating extremely heavy sedimentation. For example, early on Megahan and Kidd (1972) emphasized the problems from road construction and, in particular,

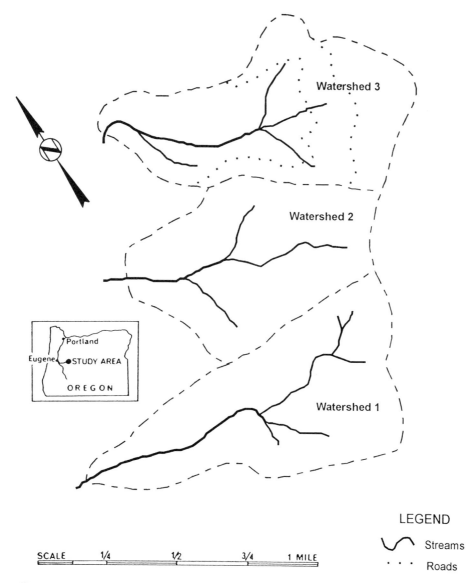

FIGURE 4.—Experimental watersheds in the H. J. Andrews Watershed Study, Oregon: Watershed 1—clear-cut with no roads; Watershed 2—control, uncut; Watershed 3—patch-cut with roads and many stream crossings (from Fredriksen 1965).

from mass soil failures associated with roads. Studying small ephemeral watersheds in the Idaho Batholith, they reported sediment production rates from logging roads 770-fold higher than rates in similar, nearby watersheds that remained undisturbed. Of that increase, 30% was due to surface erosion from roads, but 70% was due to mass soil failures. Megahan and Kidd also indicated that, although recovery was fairly rapid once logging ceased, sedimentation continued where roads had been cut through the soft, weathered granitic rock.

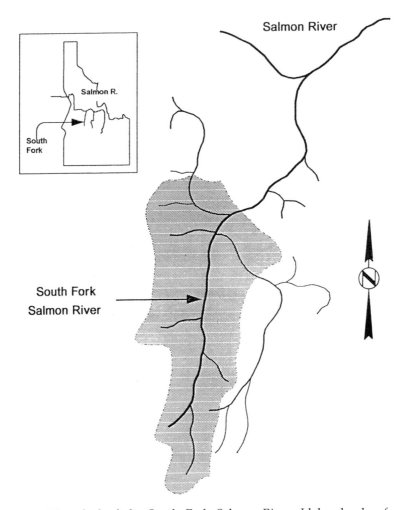

FIGURE 5.—Watershed of the South Fork Salmon River, Idaho, locale of massive streambed sedimentation and loss of salmon and steelhead trout populations (from Megahan et al. 1992).

Long-term studies of extreme sedimentation in the South Fork Salmon River (SFSR), caused by logging in the Idaho Batholith, have been documented by W. S. Platts and his associates (Platts 1968, and many subsequent reports) (Figure 5). The watershed contains valuable timber on steep hillslopes and highly erodible soils derived from the granitic bedrock. These conditions combined were thus greatly conducive to erosion and sediment production.

Platts et al. (1989) recently reviewed the history and extensive studies in the SFSR. The heavy sedimentation resulting from logging began in 1950; its severe effect on the river's salmon runs gave rise to a logging moratorium in 1965, after which sedimentation decreased. However, later sedimentation was heavy—from three major sources: (1) a massive soil movement (sand flow) in a tributary stream in 1984 that deposited huge quantities in the main stream; (2) continued erosion of a major logging road constructed near the river, by itself contributing 30% of total sediment; and (3) continuing transport of fine sediment that remained from earlier heavy deposits upstream. All of these sources apparently contributed to the long recovery lag. Seyedbagheri et al. (1987) published an extensively annotated bibliography of studies in the SFSR basin, including a short history of research and management on the river.

The highly erodible soils in the Idaho Batholith caused severe erosion from road fills. Megahan et al. (1991) studied sediment yields from three small watersheds, all crossed by a major road that had numerous fills. They emphasized that roads in forest management operations constitute the primary source of stream sediments in the granitic batholith. In this study, the authors also reported on the relative success of different preventive measures (e.g., seeding and planting) on road fills.

Other studies in the Pacific Northwest include those on spawning gravels of streams in Washington's Olympic National Park, on the Olympic peninsula. Here too, landslides associated with logging roads produced massive amounts of sediment in streams. Cederholm and Lestelle (1974) and Cederholm and Salo (1979) reported fine sediments in significantly higher quantities as the result of road-associated landslides. Similarly, Cederholm et al. (1981) and Cederholm and Reid (1987) concluded that landslides were greatly responsible for large quantities of fine sediment in spawning gravels of the Clearwater River, also on the Olympic peninsula of Washington. Cederholm et al. (1981) estimated that about half of the fine sediment was due to landslides and half to road erosion. Reid and Dunne (1984) studied sediment production as the result of the intensity of road use by logging trucks, also in the Clearwater River basin. Heavily used roads, defined as more than four loaded trucks per day, produced sediment 130-fold higher than that produced by an abandoned road, and 100-fold higher than that produced by a paved road. The heavily used road produced sediment 7.5-fold greater than that produced by the same road when it was not being used.

In another study in Washington, however, under the auspices of the Weyerhauser Company, Duncan and Ward (1985) concluded that watershed geology was more important to sediment production than forest management practices. They compared sediment from volcanic rock watersheds to sedimentary rock watersheds. More sediment was produced by the sedimentary watersheds, although some "sedimentary" watersheds had less than 50% sedimentary rock.

By the mid-1970s the effects of forest-generated sediment on salmonids had been abundantly spelled out, and the USFS in 1976 had established the Anadromous Fish Habitat Research Program (Everest and Meehan 1981). Many studies were conducted by the U.S. Forest Service at the Pacific Northwest Forest and Range Experiment Station at Portland, Oregon; Intermountain Forest and Range Experiment Station at Ogden, Utah; and the Pacific Southwest Forest and Range Experiment Station at Berkeley, California. A wealth of information on sediment in salmonid streams was generated in the following years from research conducted in this long-term program, especially from the Pacific Northwest station.

The Carnation Creek Watershed Project, Vancouver Island, British Columbia, was initiated in 1971 as a long-term research program to study the effect of

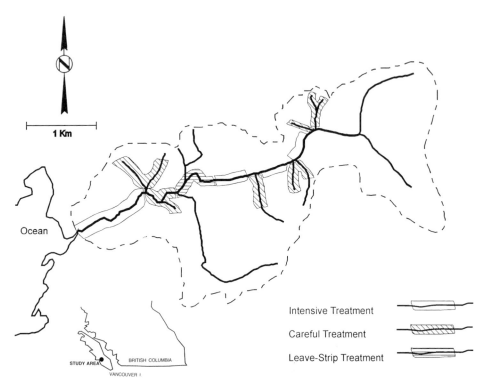

FIGURE 6.—Experimental streamside logging treatments in the Carnation Creek Watershed Project, British Columbia (from Hartman et al. 1987).

logging on anadromous fish populations (Hartman 1982; Hartman et al. 1987). Experimental logging included 5 years of prelogging study, 6 years of logging, and 5 years of postlogging study. Logging treatments included: (1) clearcutting with streamside cutting included, termed "intensive treatment," (2) "careful treatment," and (3) "leave-strip treatment" along streambanks (Figure 6).

Scrivener and Brownlee (1989) concluded that the major sediment sources in the Carnation Creek Watershed Project were: (1) landslides and debris torrents, (2) surface scour from logging roads, and (3) erosion from upstream sediment storage in bars, wedges, and streambanks during storms. Road surface erosion was not a major source of sediment in this region, because roads were constructed of blasted rock and surfaced with hard, coarse gravels. This conclusion contrasts sharply with results from the highly erodible, granitic soils in central Idaho, emphasizing the importance of matching different road construction techniques to regions of differing soil characteristics.

Elsewhere in British Columbia, studies of timber harvest practices identified skid trails and a main haul road as primary sources of sediment that reduced survival of planted rainbow trout eggs (Slaney et al. 1977b). Experiments in artificial channels produced similar results. Similarly, Roberts (1987) concluded that sediment in logged streams on the Queen Charlotte Islands, British Columbia, was mainly due to streambank erosion and landslides related to the logging. Where some British Columbia watersheds were logged to stream edges, Hogan (1986) recorded more debris torrents and thus increased sediment storage in the streambed, thereby reducing fish rearing habitat.

Sediment deposits in private forested watersheds in New Brunswick were compared between clear-cuts and controls (Welch et al. 1977). Sediment was 36% greater from clear-cut watersheds than from undisturbed watersheds; stream gravel and rock substrate were reduced and pool depths decreased in streams in the clear-cut watersheds. Logging in the clear-cut watersheds resulted in higher sediment production, the apparent result of lack of buffer strips, use of bulldozers in streambeds, and poor road design and maintenance.

In Nova Scotia, Rutherford (1986) observed the effects of 53 timber-harvest sites on Atlantic salmon streams (and revisited 47). He found siltation from initial construction at nearly 100% of the sites and concluded that logging road crossings (construction of culverts and bridges) caused most of the siltation.

The Logging Road

By far, the logging road produces the most sediment generated among forest management practices (Anderson 1971; Anderson et al. 1976; Cederholm et al. 1981; Furniss et al. 1991). Road construction also causes soil movements or landslides associated with road construction (Beschta 1978; Swanston 1991).

The density and length of logging road distribution can be major factors in determining the level of sediment production. For example, the greatest accumulation of fine sediments in streambeds occurred when the road area exceeded 2.5% of the total basin area (Cederholm et al. 1981). The authors also calculated that total road lengths of 2.5 km of road per square kilometer of the basin produced sediment more than four times natural rates. Consequently, the greatest attention to techniques to prevent sedimentation and restore damaged streams (in later sections of this volume) has been directed to the logging road.

Although the logging road has been identified as a singular sediment source, other miscellaneous sources appear in all other forestry operations that disturb the soil surface—such as log skidding, yarding (especially with tractors), site preparation, "hot burns" to remove slash (thus baring soil), heavy equipment operation, and bridging. Clear-cutting as a harvest technique has been broadly indicted many times as a major cause of eroded sediment in comparison with patch-cutting, selective cutting, and leaving buffer strips. However, all operations, it seems, must be serviced by roads. The latest summary on effects of road erosion and minimizing sediment from roads is by Furniss et al. (1991).

Finally, although the logging road has been identified as the primary sediment source—along with other sediment sources that are important in certain conditions—it should be emphasized that logging operations have many environmental effects, of which sediment production is only one. Other effects include increased amounts of woody debris that may clog stream channels and prevent fish movement; altered fish habitat in the channel; loss of riparian vegetation and resulting decreases in allochthonous inputs; elimination of streamside shade with consequent elevated water temperature; and, contrarily, an open canopy which allows more sunlight and potentially increased production by plants and invertebrates.

Whereas this volume is concerned with excess sediment as a major pollutant, several other reviews have stressed the need for more comprehensive assessment of forest management practices. Several authors (e.g., Everest et al. 1987) have pointed out that current practices in forest management have improved over the last two decades. Major changes have taken place in the Pacific Northwest, such as improved road engineering and treatment of slash. However, these authors also emphasized that sediment, although extremely important, should be integrated with total plans for forest management.

MINING

Mining operations produce immense quantities of sediments that can enter streams, elevating levels of suspended solids and turbidity, and creating deposits on streambeds. Fewer reports on sediment sources from mining appear in the biological literature than for agriculture and forestry, but local conditions have been cited as severe.

Four major forms of mining are identified: placer, surface, underground (Paone et al. 1978; Nelson et al. 1991), and sand and gravel extraction (Kanehl and Lyons 1992).

Placer mining consists of the removal of stream-transported and sorted mineral grains, often gold, from streambeds or other alluvial deposits. Surface mining includes strip mining and open-pit mining, often for coal, gold, copper, and iron ore. Underground mining uses drilled or dug shafts of various forms to extract metals and other minerals. Sand and gravel extractions are often done by dredging directly in streambeds, by excavating in stream-deposited floodplains (Kanehl and Lyons 1992), or in open-pit development in eskers and other glacial deposits removed from streams. Washing and grading operations for sand–gravel material, however, are often done using stream water, which might be returned with high sediment loads to channels.

All forms of mining can be important in the production of sediment. Two of these involve direct intrusion into streambeds or alluvial deposits: placer mining and sand–gravel extractions. However, both surface and underground mining have the potential to produce large quantities of waste materials such as spoils and tailings on the surface, which then may be subject to erosion (Hill 1975).

This volume is concerned mostly with four mining sources of sediment: placer, surface, and underground mining, and sand–gravel extractions. Stream pollution by toxic materials—sulphuric acid, metal ores, and ore-processing chemicals—can be extremely toxic to stream biota, but will not be addressed in this review. Nelson et al. (1991) provide good leads into the literature.

Placer Mining

It may be difficult to imagine the historic sourdough prospector—with his grubstake of food, gold pans, and pick-and-shovel loaded on a solitary burro, panning for tiny flakes and nuggets from a western mountain creek—having much influence on the stream's macroinvertebrates and native cutthroat trout. Panning quickly evolved, however, to the use of sluice boxes and rockers, and then—with enough capital for drilling and pumping machinery—to dredging operations that produced immense quantities of streambed deposits (James 1991; Nelson et al. 1991). Fine sediments were washed for gold and discarded in huge piles subject to erosion or, more likely, returned to a stream. Placer mining increased in efficiency, but higher technology produced even larger piles of silt, sand, and gravel to be eroded and transported into streams.

More advanced placer mining was done by two major means, dredging and hydraulic mining. Dredging consisted of digging ore-containing gravel from a suspected area, by dragline or bucket line, sorting the gravel in a washing plant, and stockpiling the tailings away from the excavation or downstream. Such tailings were immense sediment producers, but modern techniques include replacement of tailings into the excavation (USFS 1977).

A particular form of placer dredging uses suction: dredges draw up streambed material, sort, and return sediment to the stream (Griffith and Andrews 1981; Somer and Hassler 1992). Large-scale dredging produces particularly destructive effects on river channels and leaves enormous spoils that may be impossible to restore (Nelson et al. 1991).

Hydraulic gold mining soon developed to a high level of technology. A popular account of historic hydraulic mining was given by Wallace et al. (1976), wherein water pumped from creeks and rivers, sometimes miles away, was aimed in water jets at banks and hillsides suspected of containing rich quantities of gold grains. Whole hills were leveled to provide the materials subsequently processed by water sorting. The sediment transported downstream was enormous, although in those early days effects on the stream biota were of no interest. Between 1853 and 1884, "hydraulicking" generated high sediment loads in streams of California, including 270 million cubic yards in the Bear River basin (a tributary of the Feather River), second only to that in the much larger Yuba River basin (James 1991). This form of mining was enjoined in 1884, but by 1950 the Bear River had not returned to pre-mining conditions. Some of the early reports of damage to stream fisheries concerned hydraulic gold mining (Smith 1940; Shaw and Maga 1943). Few deposits remain today that could be economically mined in this way under current legal restraints (USFS 1977).

Operations in Alaska that extract placer gold deposits by stripping are important economically and politically (Wagener and LaPerriere 1985). The deposits are found at the interface of alluvial gravels and bedrock; the overburden is stripped to reach the gold-bearing gravel. Huge amounts are thus stripped, leaving both streams and streambanks in greatly disturbed conditions. The gravel is then sluiced to separate the gold. The washwater is often returned to the stream, where high concentrations of suspended sediment result, to the great detriment of primary producers, invertebrates, and fish populations (Birtwell et al. 1984; Wagener and LaPerriere 1985; Van Nieuwenhuyse and LaPerriere 1986; Pain 1987). The sediment source is largely twofold: erosion from the spoil piles and streambanks and the finer particles, especially clay, in the washwater that escapes settling and filtering operations and moves downstream. The results are severe suspended sediment conditions, high turbidity, and cement-like blankets of deposited sediment on streambeds.

Hydraulic mining for gold produced immense quantities of water-sorted gravel and sediment, historically deposited in stream channels, and great spoil piles remain. Central Idaho. (Photograph courtesy of William S. Platts.)

Surface Mining

Surface mining has been labeled the greatest potential source of sediment generation among all forms of mining. The main source of sediment is in the erosion of disturbed land left without vegetative cover. In a review of eastern United States mining, Starnes (1985) concluded that sedimentation " . . . is a universal problem and becomes especially severe where multiple mines are present in the same watershed and the assimilative capacity of the receiving streams is exceeded." Nelson et al. (1991) state that " . . . erosion of surface-mined lands and mine spoils represents one of the greatest threats to salmonid habitat in the western USA." Richardson and Pratt (1980) provided an extensive annotated bibliography on environmental effects of surface mining.

Mining by excavating land surfaces takes several forms, but the major operations may be described in two categories: strip mining and open-pit mining. Strip mining involves following a mineral seam or stratum, with overburden wastes deposited in spoil piles or banks near the excavation. Extraction of coal is the most typical example of strip mining. Spoil piles may be large, including silt- and sand-sized particles subject to erosion and transport to stream channels (Musser 1963; Spaulding and Ogden 1968).

Most strip mining has been for coal and has occurred in three major North American regions: (1) Appalachia, including mainly Virginia, West Virginia, Kentucky, and Tennessee; (2) Interior, including mainly Illinois, Indiana, and Missouri; and (3) West, the Rocky Mountains and their east slope in both the United States and Canada (Vories 1976; Thames 1977; Schaller and Sutton 1978; Thirgood 1978; Starnes 1985). Since the U.S. Surface Mining Control and Reclamation Act (SMCRA) of 1977 (30 U.S.C. §§ 1201 to 1328), mining and reclamation operations have been consistent in the United States; SMCRA requires that all water from mining-disturbed lands must pass through sediment-filtering structures (Bell and Payne 1993). However, immense abandoned spoils remain from previous mining. The major sediment source today is in abandoned coal mine spoils, totalling 1.7 million acres in the United States; the largest amount of sediment is generated on the steep hillslopes of the Appalachian region (Starnes 1985).

Strip mining for coal generates the most erodible spoils. It takes three forms: contour mining, area mining, and mountain-top removal. Contour mining follows along a coal seam on a slope; the spoils are deposited along the outer edge, often to slide downslope on the mountain side. Area mining is practiced largely in the Interior region on flatter topography, where spoils are deposited in ridges, leaving a "washboard" aspect. Mountain-top removal includes the complete removal of a peak, leaving a flat top with spoils deposited around the outer edge. Contour and mountain-top mining is most common in the Appalachian region, where erosion and sediment generation are the greatest (Starnes 1985).

Coal mining is the largest single contributor of surface-mined spoils. Although the production of acid waste (from sulfur-containing iron pyrite, associated with some coal deposits) is of great concern as a toxic pollutant, the most important mining source of sediment pollution is the generation of sediment in the overburden material deposited in spoil piles from coal mines.

Open pits involve a relatively discrete excavation, often large in area, and

Surface mining leaves huge excavated pits and spoil piles. Such voids are often allowed to fill with water and may be managed for sport or commercial fisheries. Iron mine in Mesabi Range, northern Minnesota.

sometimes exceedingly deep. Open-pit mines are frequently used for copper, phosphate, and iron ore, as well as coal extraction. Open-pit mining produces large accumulations of waste materials as sources of sediment, usually at the mill site where as much as 98–99% of the ore becomes mill tailings. These wastes are often disposed of into tailings ponds or trenches constructed to contain a large area that will build slowly in elevation. The ponds are protected by dams or embankments that hold back drainage. Water is expected to evaporate or drain and filter underground. Sediment problems arise when the tailings pond over-flows due to a flood or when the dam fails due to poor design. Duchrow (1982) described the effects of the failure of a dam holding back red-clay tailings from a barite mining operation in Missouri. Turbidity increased in two Ozark streams, killing fish and reducing the benthos. The abandoned tailings pond, especially when poorly designed in the first place and poorly maintained subsequently, is a major environmental problem in production of sediment (USFS 1977).

Underground Mining

Underground mining involves digging or drilling tunnels, from which ore is extracted mechanically from tunnel walls. The tunnels may be in the form of a shaft (vertical), a slope (inclined), or an adit (horizontal). The amount of waste produced depends upon the kind of mineral—metalliferous ores produce a high proportion of waste, whereas nonmetallic deposits generate smaller quantities (Nelson et al. 1991). At mills that process ores from underground mines, the disposal of waste tailings and their sediment-producing potential is similar to that from surface mines. Tailings ponds, constructed near the processing mill, present problems similar to those of spoil piles from open-pit metalliferous mines, where they constitute potential sediment sources.

The release of toxic chemicals from surface and underground mining wastes, particularly metals and sulphuric acid, has degraded surface waters on a broad scale. Toxic materials often remain as precipitates or are adsorbed on sediment particles stored in a stream system, to be released later into an active state (Nelson et al. 1991). The problem of such contaminated sediments, although not included in this volume, is of rapidly growing concern and has been the subject of extensive publication elsewhere (e.g., Baudo et al. 1990).

Clearly, excess sediment from mining operations constitutes an extreme source of stream pollution, particularly threatening salmonid fisheries in the western United States.

Sand and Gravel Extraction

Mining of sand and gravel takes two major forms: instream dredging of the streambed, and floodplain excavations (or wet pits) that often involve connecting ponds and an outlet to a stream (Newport and Moyer 1974; Blauch 1978; Kanehl and Lyons 1992). Starnes (1983) pointed out that sand and gravel are probably the most commonly mined resources of all, but, ironically, the industry is the least regulated of all mining activities.

Kanehl and Lyons (1992) and Nelson (1993) listed several physical effects on streams from sand and gravel mining, most of which may exacerbate sediment entrainment in the channel: changed channel morphology, increased velocity, headcutting (upstream eroding action due to increased velocity), and streambed modification. Some or all of these results may include increases in fine-particle deposits (deposited sediment) and elevated turbidity from suspended sediments. The authors included a survey of stream studies from six states and reported reduced photosynthesis, decreases in invertebrate biomass, and decreases in biomass and number of fish species.

Recovery from damage caused by sand and gravel mining may be exceedingly slow, with larger silt and sand deposits slow to dissipate. Kanehl and Lyons (1992) found conditions in the Big Rib River, Wisconsin, to remain in an early stage of recovery (mainly a sand–gravel substrate and few pools and riffles) 20 years after mining had stopped. Stream reaches 10 years after mining were in worse condition, with sand the predominant substrate, no riffles or pools, wide channels, unstable streambanks, and no fish cover. A section below active mining operations had a bottom substrate of almost total sand. Nelson (1993) also pointed out that instream dredging and gravel washing produce fine sediments in all flow conditions, so that in low discharge, fine particles settle in riffles as well as in all other habitats. Sedimentation on the streambed resulting from such operations often contains higher proportions of fine particles.

Gravel mining operations on Arctic and subarctic floodplains in Alaska produced severe physical effects on streams, including channel alteration and creation of ponds that attracted fish in summer and froze solid in winter. Effects on fish populations included elimination or reduction of populations of Arctic char, Arctic grayling, and other fish species (Woodward-Clyde Consultants 1980a).

Related to sand and gravel extraction is quarrying for removal of construction-quality rock, such as granite and sandstone. This surface extraction probably would not produce waste tailings in the same quantities generated by open-pit or strip mining, because most of the material removed is used. No references were found in the literature on effects of sediment originating from quarry wastes.

In his early paper on fish distribution in Colorado and Utah, Jordan (1889) decried the siltation of mountain streams by placer mining and the loss of native fishes, perhaps the first published statement on the deleterious effects of sediment waste from mining. Subjects on mining effects cited later include: hydraulic gold mining (Sumner and Smith 1940); suspended sediment effects on Arctic grayling in Alaska and Yukon streams (LaPerriere et al. 1983; Wagener and LaPerriere 1985; McLeay et al. 1987); suction gold dredging (Thomas 1985; Harvey 1986); surface and strip mining (Boccardy and Spaulding 1968; Spaulding and Ogden 1968; Branson and Batch 1972); gravel dredging (Forshage and Carter 1974); and gravel washing (Cordone and Pennoyer 1960). Many other reports contain brief information on forms of mining as sources of sediment.

URBAN DEVELOPMENT

Urban development can be extremely important as a source of anthropogenic sediment in metropolitan areas (Wolman 1964; Simmons 1976). Compared with agriculture, forestry, and mining, however, it appears to be less extensive in geographical distribution. Residential development, industrial construction, installation of streets and utilities, and all other elements of the infrastructure required in densely populated locales can concentrate sediments. Any kind of excavation—earth moving, drainage, bridging, tunneling, or other activity that disturbs soil surfaces—can serve as a source of soil loss and sediment transport to nearby streams.

Most concerns and published research results, both informal and primary, involve residential areas in Maryland, near developing urban and suburban Baltimore and Washington, DC (Wark and Keller 1963; Wolman 1964; Wolman and Schick 1967; Guy and Ferguson 1970; Randall et al. 1978; Yorke and Herb 1978). The early survey by Wark and Keller (1963) included the analysis of 42 stream locations and measurement of total sediment loads (suspended load plus 10% for bed load). Sediment loads from lands undergoing urbanization were up to 50 times more than those in rural areas. Wolman and Schick (1967) measured annual sediment yields from urban development to streams and found up to 200 tons/acre or more, far more than from agricultural erosion in the same area (about 5 tons/acre); stream sedimentation blanketed benthic faunas and altered fish species composition. A major problem was the period of time that disturbed surfaces lay exposed—more than a year at 25% of the sites.

The Occoquan River basin lies on the southwestern periphery of the heavily urbanized Washington metropolitan area. Uniquely, the river had a rapidly expanding urban area along an intermediate upstream reach and an important water supply reservoir downstream at the river mouth in which water quality had

Mountain roads to service recreational development—in this case, in the Blue Ridge region of North Carolina—can produce severe sediment transport downslope to streams. The mass soil movement on the right of the photograph is a common result in such topography.

rapidly declined. Randall et al. (1978) studied these water quality problems and determined that suspended sediment was the greatest single pollutant; organic pollution and heavy metals were also important.

An evaluation of control practices was undertaken over the period 1962–1974 in the Anacostia River basin, north of Washington, DC, where yields of suspended sediment from urban land were up to six times more than yields from cultivated land and up to over 120 times those from forested and grassed lands (Yorke and Herb 1978). Control measures, including shorter exposure times, installation of settling basins, temporary vegetation during construction, and earlier final vegetation installations, decreased suspended sediment yields by 60% to 80%.

In a land-use study in Virginia, Jones and Holmes (1985) compared urban, agricultural, and forested areas, summarizing effects on state water resources. They concluded that forestry practices contributed little, agriculture was an important source, and urban development contributed the most sediment (as well as other pollutants).

Faye et al. (1980) compared erosion and suspended sediment yields in nine watersheds in the Upper Chattahoochee River basin, Georgia, and reported the greatest suspended sediment yields from urban areas, compared with forested and agricultural lands. Krug and Goddard (1986) reported on studies in Wisconsin, where urbanization of the basin of Pheasant Branch, a tributary of Lake Mendota, had caused sediment loads to be discharged into the lake that required dredging to maintain recreational boating. Sediment-reduction structures were installed and studied by the U.S. Geological Survey in an effort to plan for the effect of continually increasing urbanization in that area (Krug and Goddard 1986).

A comprehensive handbook for best management practices (BMPs) prepared by the Wisconsin Department of Natural Resources and Wisconsin League of Municipalities describes many practices in urban development that may generate excess sediment, as well as techniques to curb erosion around construction sites (WDNR 1992). One of the most important factors cited was the length of time that disturbed lands were left exposed to weather and precipitation.

STREAMBANK EROSION

The erosion of streambanks is a natural process that takes place under normal conditions in streams. Lateral migration of stream channels across floodplains is readily observable in stream valleys unaffected by human activities. However, cultural activities that affect the amplitude of discharge fluctuations can exacerbate erosion of streambanks. Other land uses that cause or accelerate destabilization of streambanks, such as channelization, trampling by livestock, and recreational traffic can increase the delivery of sediment. In these cases, the increased sedimentation can be considered anthropogenic. Whether the sediment produced by streambank erosion is natural or anthropogenic, however, the quantities involved may be large and are often greatly detrimental to water quality and fisheries management.

Two processes are chiefly responsible for streambank erosion: (1) fluvial entrainment of bank material by high discharges, and (2) mechanical failure of the bank, allowing material to slump to the basal area, where normal discharges can entrain the added sediment into the stream flow. Bank failure is influenced by cohesiveness of bank material, moisture content, and slope of the bank. Removal of slumped material at the basal area by high water currents increases the slope and eventually creates conditions that cause further slumping. Alonso and Combs (1990) illustrated three forms of bank failure: (1) rotational slip, where the failed mass slips in a downward and inward rotation toward the stream; (2) plane slip, initiated by surface cracks in the bank, where the direction of slip is straight but angled toward the stream; and (3) cantilever failure, where an undercut bank falls into the stream (Figure 7). The authors also pointed out that a deepened channel bed, due to any factor, causes increased height and slope of the bank, leading to further instability and greater likelihood of failure. Thorne (1982) discussed bank erosion from an engineering standpoint.

Striffler (1964) sampled sediment and stream discharge in relation to land use at 20 stream sites in Michigan's Lower Peninsula. Streams in watersheds of forest and wild land (brush, swamp, other well-vegetated idle land) generally had stable flows and less sediment; streams in cultivated and pastured watersheds had heavy sediment deposits and variable flows. Streambank erosion was an important factor in the streams with high sediment loads. Lines et al. (1979) estimated the extreme sediment losses from streambanks (10 million cubic yards per year) in the lower reaches of several western Oregon and Washington rivers, resulting from encroachment toward streambanks by agricultural and urban development in upstream reaches.

Some naturally eroding streambanks are virtually uncontrollable. The Nemadji River, located in Minnesota and Wisconsin, flows over glacial lake clay deposits several hundred feet deep. The thick clay was deposited under an early stage of glacial lakes in the Lake Superior basin during glacial melting about 10,000 years ago. Eroding banks leave the river almost perpetually turbid.

Stream straightening and channelization may be short-term solutions to flooding of agricultural fields, but in the long term such practices increase flow velocity at sites of outside bends in previous meanders; the result is often severe streambank erosion and high sediment inputs. Urban streams may be more subject to streambank erosion because of cultural requirements that they be straightened and narrowed (Wilson 1983). The practice of channelization and drainage to increase arable land leads similarly to increased streambank erosion, due to the tendency of streams toward meandering and accelerated hydraulic action on streambanks (Nunnally 1978; Fajen 1981).

In their study of fish populations in Court Creek, Illinois, Roseboom and Russell (1985) noted that large game fish were eliminated in channelized stream segments that were heavily sedimented as a result of streambank erosion. The development of gullies in an agricultural terrain can also be a major source of sediment. Channelization and stream straightening steepen the gradient and raise water velocities which tremendously accelerate erosion of the streambed through incision of the channel (Harvey et al. 1985).

Hansen (1971) reported an exceptional case of streambank erosion in the Pine River, Michigan, where storms impacting high, eroding banks of clay and sand produced dense clay turbidities and rapidly eroding banks. A 640 acre-foot reservoir built for electric power generation was completely filled in a 40-year

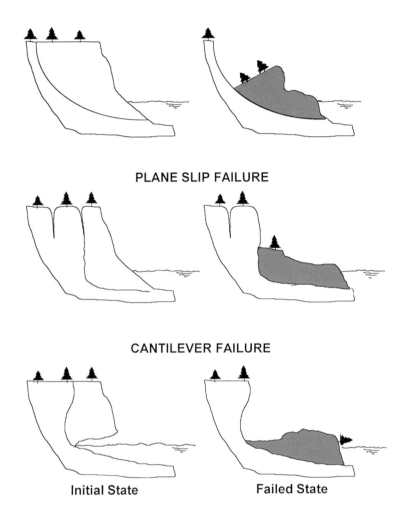

PLANE SLIP FAILURE

CANTILEVER FAILURE

Initial State   Failed State

FIGURE 7.—Types of streambank failure in which the slumped bank material becomes entrained by stream currents (from Alonso and Combs 1990).

period. Data in a 17-year record suggested an increasing average discharge. Although earlier discharge data apparently were not available, there is the further possibility of previous increased discharges, in view of past logging of white pines, with the reduction of transpiration losses that the forest change probably brought about.

Logging practices can have a great impact on streambank stability. For example, past logging practices on watersheds in the Queen Charlotte Islands, British Columbia, resulted in wider, shallow channels and heavy sediment deposition, the result of tree-cutting near stream edges (Roberts 1987). Similarly, historic log drives in the 1800s in Michigan's present Hiawatha National Forest area produced heavy streambank erosion. The major effect of these river drives

Channelized streams and ditches, common in the flat topography of many agricultural regions, have a natural tendency to meander, reinitiating a winding stream by eroding into floodplains. Because the natural stream is shortened, gradients are increased, and incision is accelerated to erode the channel bed.

was the disruption of streambanks which were composed largely of glacially deposited sand, resulting in long-lasting sand deposits on streambeds and reduction of fish habitat (Bassett 1988). An intense restoration program was planned, including designs for subsequent streambank control.

In their review of stream fish habitat in the Midwest, Lyons and Courtney (1990) concluded that streambank erosion was a major cause of sediment production in warmwater streams. In most cases, warmwater streambank erosion was linked to agriculture: susceptible streams on floodplains that were historically attractive to settlement and agriculture, loss of riparian zones due to intense cultivation, and trampling by livestock.

Streambank erosion is a major contributing factor in sediment generation where it occurs in floodplains that contain previously stored sediment. Human activity in cultivated, grazed, forested, or urban regions has the potential to initiate or exacerbate streambank erosion. Trimble (1983) reported that lateral (streambank) erosion has been the major form of soil removal from floodplains since European settlement. Cultivation of row crops too close to the streambank removes protective vegetation; livestock grazing near stream edges will cause trampling and bank slumping; timber harvest along streams removes cover and promotes bank failures. Like roads that must service almost any kind of development or extractive activity, the streambank is a common element that may erode and contribute to some of the most severe cases of stream sedimentation.

MISCELLANEOUS SOURCES

Almost any form of land disturbance has the potential for producing excessive sediment to be transported to streams. Naturally occurring sediment movements may result from landslides or extraordinary storms, uncontrollable by human intervention (Coats et al. 1985). Overall, however, most sedimentation is the product of human activity. Even what may be seen as natural erosion may have origins in historic, human-caused landscape modifications.

Many roads and highways, built not for resource extraction but for public transportation, can also generate eroded sediment. Vice et al. (1969) sampled sediment from a basin tributary to the Potomac River, Virginia, for 3 years, during which time the basin was under intensive highway construction. The

Bridge and highway construction at river crossings involve large quantities of sediment, a potential source of stream sedimentation. Some time prior to this photograph, an exceptional storm moved great quantities of sediment, blocking the highway and emptying huge amounts into the river. St. Croix Valley, Wisconsin.

construction area, although occupying only 1–10% of the basin area, contributed 85% of the sediment. Almost all sediment movement occurred during storms. Wright et al. (1978) cited the importance of sediment eroded from cut, fill, and median slopes during highway construction and listed means of erosion control. Chisholm and Downs (1978) reported that construction of a four-lane highway along Turtle Creek in West Virginia generated massive sediment, burying the streambed under 10 in of deposits and eliminating the stream benthos. After construction ceased, however, subsequent rains scoured the deposits downstream and returned the stream to its previous condition—including the return of benthic invertebrate populations recolonized by drift from an undisturbed tributary.

Erosion from footpaths and recreational trails for foot, horse, or motorized traffic can have localized but severe effects. Helgath (1975) reported erosion from trails and sediment movement in the Selway-Bitterroot Wilderness, Idaho and Montana, due primarily to trail location and other terrain characteristics (Figure 8).

The effect of extreme events, even though rare, can be severe and long-lasting. Coats et al. (1985) reported results of an exceptional rainstorm in the watershed of the San Lorenzo River, California. The storm caused unusually heavy deposits of sand that filled pools and covered riffles in the river. The sources of sand were mainly roads and landslides.

Activities associated with road crossings can accelerate inputs of fine sediment. Such activities include bridge construction, culvert installation, and diversions with coffer dams. Stream crossings studied in Virginia resulted in large reductions in the number of species and abundance of both fish and invertebrates (Reed 1977). In these stream crossings, downstream drifting of benthic organisms from the construction areas, due to high suspended sediment concentrations—rather than heavy deposits—appeared to be the responsible factor. Highway

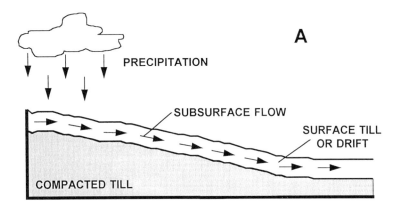

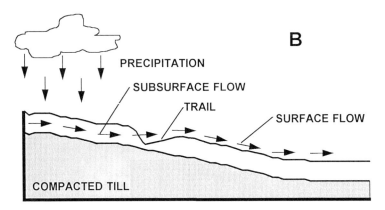

FIGURE 8.—Surface erosion resulting from development of recreational trails. Construction and use of trail intercepts subsurface flow, causing it to flow above ground. (A) Subsurface flow before trail construction; (B) Subsurface flow interrupted and diverted to surface by trail construction (from Helgath 1975).

construction along Joe Wright Creek, a mountain stream in Colorado, reduced macroinvertebrates where sediment deposits occurred, but subsequent high streamflows rapidly returned the stream benthos to its natural condition (Cline et al. 1983).

It appears that most such short-term construction projects, done essentially at a point on a stream, will have temporary effects, as opposed to forest roads that have effects over long distances and times. If, at such stream crossings, subsequent flows are high enough to scour away light deposits, invertebrates and fish generally will repopulate quickly when undisturbed upstream areas contain fish, and invertebrates remain available for recolonization by drift. Conversely, low discharge, shallow gradients, and large sediment deposits may have much more persistent effects. However, the fate of sediments washed farther downstream, and the length of time required for complete removal or assimilation, have seldom been evaluated.

The construction of electric transmission lines in mountain regions requires special treatment of tower sites, here including slope reduction, grass plantings, and temporary plastic sediment fences. Tellico Valley, North Carolina.

Prescribed fire, most common in forest management, can produce serious sediment losses from affected surfaces. The principal factors are the exposure of mineral soil surfaces left devoid of the protection of vegetative and forest litter cover, and steepness of hillslopes. Hansen (1991) measured soil loss in plots of varied steepness, following skyline logging by overhead cable (no skid roads) and burning to remove logging debris and to reduce hardwood competition. He reported annual soil losses up to 3.75 tons/acre on 30% slopes, 5.89 tons/acre on 50% slopes, and 18.68 tons/acre on 70% slopes.

Experimental burning on sagebrush rangelands produced sediment yields that varied with fire intensity (Simanton et al. 1990). Plots with low-intensity burns yielded two times more sediment than unburned plots; high-intensity burns yielded five times more sediment than unburned areas. In recommending fall burning, the authors pointed out that burning in spring followed by heavy precipitation produces much more sediment than does burning in the fall, when rainfall is less. Emmerich and Cox (1992) simulated rainfall on experimental burn plots in two areas of southeastern Arizona; they observed no significant increase in runoff and sediment from burned plots but surface slopes varied only between 5 and 7%.

The principal recent review of fire-effected erosion, with an extensive bibliography, is that by McNabb and Swanson (1990); they concluded that the degree of erosion depends upon the severity of the fire, soil erodibility, slope steepness, and subsequent precipitation. Other results of fire include increased stream flow due to reduction of transpiration losses and increased likelihood of mass soil wasting after plant roots had decayed in the soil. Such increases in soil loss rates may persist for several decades. Erosion caused by fire may total 25% of all forest sediment losses. However, McNabb and Swanson (1990) also concluded that generally, fire-related sediment production on forest lands is not serious when compared with sediment production from other forest practices, such as certain timber-harvest methods and roads.

Other operations that can generate sediment, less common but potentially important locally, include: sediment flushing from silted-in reservoirs used for electric power generation, with severe downstream effects on fish and invertebrates (Hesse and Newcomb 1982); electric transmission rights-of-way across streams, such as rock fords and culverts (Peterson 1990); electric transmission line crossings in a trench under a stream (Cerretani 1991); pipeline crossing by trenching (Brunskill et al. 1973; Tsui and McCart 1981); and backwash from a water-filtering facility (Erman and Ligon 1988). Habitat alteration (stream improvement) for fisheries enhancement can also generate sediment if the construction encroaches upon the stream channel, possible changes in stream morphology are not evaluated, and appropriate protective measures are not included (Apmann and Otis 1965).

EFFECTS OF
SUSPENDED SEDIMENT

Suspended sediment, usually particles of silt and clay (about 2 to 60 μm), is borne by normal stream currents at nearly the same transport velocity as the water. Various sizes of sand may be carried in suspension under conditions of high current velocities. Because much of suspended sediment is composed of clay particles, which are adhesive to some degree, effects upon stream biota include coating of biologically active surfaces, both plants and animals.

Suspended sediment can abrade and suffocate periphyton and macrophytes, as well as decrease primary production (photosynthesis rate) because of light reduction. It can disrupt respiration and modify behavior of invertebrates. In fishes, respiratory capacity of gill surfaces may be lost and vision and feeding efficiency diminished. Eggs and fry in redds or stony interstitial spaces may be suffocated, and thus reproductive success may be reduced. Migrating fish will avoid streams with high suspended sediment concentrations.

A distinction is made between suspended sediment particles in water entering a salmonid redd, for example, which may coat eggs and larvae, and the complete covering of the overlying streambed by deposited sediment, such as heavy sand or clay deposits. The effect of sediment on fish reproductive success is the subject that has attracted the greatest research attention by aquatic biologists concerned with sediment in streams, probably due to the great economic interest in the salmons and trouts. A large proportion of the literature on biological effects of stream sediments concerns effects on fish eggs and fry, particularly in salmonid redds.

Suspended sediments contribute to turbidity and thus affect light transmission through the water and to the streambed. Effects of turbidity on stream biota are related to, but sometimes not clearly distinguished from, the direct effects of suspended sediment. Precisely defined, turbidity is an optical property of water resulting in a loss of light transmission due to absorption or scattering (Dieter 1990). Whereas suspended sediment may be the major contributor to turbidity, other materials (e.g., colloidal color, plankton, organic detritus) also contribute to the reduction of transmitted light.

Turbidity is most commonly measured in the field with a nephelometer in nephelometric turbidity units (NTU) (APHA et al. 1992). On the other hand, suspended sediment is measured in units of parts per million (ppm; mg/L) and samples are usually transported back to the laboratory for analysis. Because

The single-stage sediment sampler used to collect water samples for suspended sediment analysis during all stages of water discharge fluctuation. The concentration of suspended sediment is of primary concern to fish reproductive success when downwelling stream water is the source of water to salmonid redds. Sampler is described by U.S. Army Corps of Engineers (1961).

suspended sediment in many situations may be the sole contributor to turbidity, the technique of using a nephelometer to measure turbidity has sometimes been substituted for the actual measurement of suspended sediment concentration, using an empirical relationship to approximate suspended sediment concentration from NTU values (Kunkle and Comer 1971; Earhart 1984). Many regression equations have been published for this purpose (Truhlar 1976; Sigler et al. 1984; Lloyd et al. 1987; Scannell 1988). However, the relationship may vary considerably with other sources of turbidity that may be present and also with the size of suspended sediment particles. Caution should be used in any attempt to substitute turbidity measurement for direct measurement of suspended sediment concentration (Everhart and Duchrow 1970; many later authors).

Under storm conditions, suspended sediment concentrations commonly reach thousands of parts per million (rarely, over 100,000 mg/L, Lloyd 1987; Braatz 1993a), but increases in turbidity of only 5 NTU may have deleterious effects upon aquatic organisms. Some states have maximum turbidity standards of 5 to 25 NTU above natural conditions, which may translate to only about 3–7 mg/L of suspended sediment (Lloyd 1987).

A group of scientists from several countries made an early attempt to develop suspended sediment criteria for European freshwater fisheries in lakes and streams (EIFAC 1965). Their extensive literature search on effects of sus-pended sediment on fish included direct effects such as mortality and disease; sublethal effects on reproductive success, growth, behavior, food supply (macro-invertebrates); and effects on fisheries (fish populations and harvest). Their ten-tative criteria for continuous suspended sediment were: suspended sediment less than 25 mg/L, not harmful; suspended sediment 25–80 mg/L, fish yield some-what reduced; suspended sediment 80–400 mg/L, good fisheries unlikely; and suspended sediment more than 400 mg/L, poor fisheries. Although the literature cited was limited, especially on North American streams, this publication consti-tuted the first major attempt to intensively review the biological research on suspended sediments.

Other reviews have dealt in whole or in part with the direct effect of suspended sediment, or indirectly with the effects of turbidity caused by sus-pended sediment. Wallen (1951), conducted some of the earliest experiments with suspended sediment and warmwater fishes. Everhart and Duchrow (1970) pre-pared a generalized treatment for the U.S. Bureau of Reclamation, and Gammon (1970) contributed an early USEPA report. Hynes (1973) briefly summarized sediment effects on stream biota; his review appeared in a symposium largely devoted to hydrology and fluvial processes. Sorensen et al. (1977) provided a large USEPA report as a basis for development of standards of suspended sediment and dissolved solids in freshwater environments. Farnworth et al. (1979) published another large USEPA report (on nutrients as well as sediments) including a chapter on inorganic and organic sediment in surface waters. Another large USEPA report, by Muncy et al. (1979), has been the only review dealing with sediment and warmwater fish reproduction. Petticord (1980) contributed a chap-ter to a book mainly on contaminated sediments but included a brief summary of suspended sediment and aquatic organisms. Lloyd (1985, 1987) reviewed many papers on the effect of turbidity on all components of stream communities, in an attempt to develop water quality standards for Alaska waters. Dieter (1990) developed an annotated bibliography on turbidity, dealing mostly with lentic waters. Newcombe and MacDonald (1991) recently reviewed suspended sedi-ment effects in fresh and marine waters, with emphasis on direct effects on salmonid fishes and some invertebrates.

EFFECTS ON PRIMARY PRODUCERS

Reduction of water transparency caused by suspended silt and consequently, reduced light available for photosynthesis were main points emphasized by Ellis (1936) in his early-warning paper. He gave no data nor did he quote data from others. Ellis also expressed concern for possible effects of suspended sediment on other water characteristics, such as temperature and conductance.

Subsequent work quoted by Cordone and Kelley (1961) on effects of sus-pended sediment included very little on the possibility that suspended sediment severely affects plants. They also reviewed published work involving effects on photosynthesis, but their discussion comprised mostly speculation or quotations of other authors' speculative discussions. No literature citations were made from primary research journals (except Ellis 1936), but Cordone and Kelley cited several unpublished reports wherein it was concluded that suspended sediment reduced photosynthesis. They also encouraged increased research on the effects of turbidity on algae.

Among works that attempted to relate turbidity measurements (in NTU) to estimates of gross primary production (e.g., in grams oxygen per square meter per day [g O_2/m^2 per day]), the work of J. D. LaPerriere and her associates on the effects of gold mining on Alaska streams stands out (LaPerriere et al. 1983; Van Nieuwenhuyse and LaPerriere 1986; Lloyd et al. 1987; Pain 1987). Measuring turbidity, gross primary production, and other variables, they found a negative correlation between turbidity caused by suspended sediment—from placer gold mining—and primary production. In unmined streams with turbidity at 1 NTU, daily gross primary production was 0.2–1.2 g O_2/m^2; in moderately mined streams with turbidity at about 170 NTU, gross primary production was reduced to 0.08–0.64 g O_2/m^2 per day; and in heavily mined streams with turbidities between 1,100 and 3,400 NTU, primary production was undetectable and chlorophyll measurements were zero. Heavy deposits of clay from suspended sediment released from gold-ore sluicing in some cases completely eliminated periphytic algae on the streambed (Van Nieuwenhuyse and LaPerriere 1986; Pain 1987).

Lloyd et al. (1987) developed a model relating turbidity (in NTU) to gross primary production. An exemplary result was that in a clear, shallow stream an increase in turbidity of 5 NTU would decrease gross primary production by 3–13%, and an increase of 25 NTU would decrease gross primary production by 13–50%. (Effects of suspended sediment upon benthic macroinvertebrates and fish in the same streams were similarly detrimental and are discussed in a later section.)

Increased turbidity due to gravel extraction in a French river decreased diatom density, chlorophyll, and primary productivity (Rivier and Seguier 1985). Sloane-Richey et al. (1981) and Graham (1990) pointed out that clay-sized inorganic particles may pervade algal cells in periphyton by adhering to sticky surfaces, thus reducing the organic proportion in periphyton and therefore, the food value of epilithic algal cells for grazing invertebrates. For example, in a clear stream, with suspended sediment at 1 g/m^3 (1 mg/L), 52% of the periphyton dry weight was organic, but at suspended sediment levels of 2–5 g/m^3, only 22% of the periphyton dry weight was organic (Graham 1990).

A possibly confounding result occurs in the case of forest clear-cutting near streams: an increase in light availability with the removal of riparian vegetation. Much concern has been expressed in the literature about increased sediment input from logged areas (see later sections), but clear-cutting near streams may also increase primary production with the increase of light (Burns 1972; Lyford et al. 1975; Hall and Ice 1981; Murphy and Hall 1981; Webster et al. 1983; Adamus et al. 1986; Johnson et al. 1986).

Some authors have argued that the effect of increased light on the stream surface may overcome the effect of increased suspended sediment concentration and turbidity that might simultaneously occur. However, an increase of sediment inputs, creating excess deposited sediment which also results from riparian disturbance, is likely to cause instability of the stream substrate and a resulting decrease in the abundance of aquatic insects (Gregory et al. 1987). Saturation of benthic algal photosynthesis occurs at about 20% of full sunlight; consequently the authors urged caution in assuming that more sunlight will necessarily increase primary production. They pointed out that above the 20% saturation level the increased primary production—due to canopy removal—may not compensate for the reduction in overall productivity caused by turbidity and substrate sedimentation.

Little is known about the scouring effect of suspended sediment upon attached or benthic primary producers. In laboratory stream experiments, Horner et al. (1990) determined that suspended sediment additions to their streams caused loss of periphyton cells only when current velocities were high, approximating those velocities that would be expected in spates or floods.

So far, there is little evidence that suspended sediment—even at high levels—will damage stream communities through reduced photosynthetic rates, especially if high suspended sediment levels are not extended over a long term. It seems likely that high levels of suspended sediment would have direct effects on other trophic levels (macroinvertebrates, fish) that might overwhelm the positive effect on primary producers resulting from canopy removal.

Within the framework of the river continuum concept (Vannote et al. 1980), the lowermost reaches of large rivers (approximately 7–10 stream orders) may be permanently turbid owing to suspended sediment and organic seston and, therefore, totally without photosynthesis. In these conditions, secondary producers such as benthic invertebrates depend chiefly upon drifting fine organic particles for food. On the other hand, in headwaters and low-order (1–3) tributaries, stream bottoms may be shaded so completely by a riparian canopy that sunlight required for photosynthesis is totally occluded. In these conditions, the source of available primary or vegetative matter is terrestrial (i.e., allochthonous organic matter) and is processed by decomposers and invertebrates in the stream to sustain often high rates of secondary production. In neither of these two conditions—headwater streams nor large rivers—would the absence of light for photosynthesis appear to be a significant factor affecting higher trophic levels.

In middle reaches (approximately stream orders 4–6), where the riparian canopy does not extend over the stream and where low turbidity normally permits sunlight to reach stream bottoms, energy is supplied to the stream through photosynthesis by benthic plants. In these conditions, high suspended sediment concentrations and turbidity, especially if extended over a long time, may have a negative effect on secondary productivity through the loss of primary production. However, there is little in the sediment literature that relates suspended sediment to reduced primary production and, in turn, to reduced secondary production (e.g., by macroinvertebrates) because of the loss of their food resource. Further research on this inter-community problem would be a fruitful approach.

There is little information in the literature dealing with the effects of deposited sediment on primary producers. Only speculation about such effects was included in some early papers that did not distinguish between suspended sediment and deposited sediment; no specific data have been published. Probably deposited sediment heavy enough to inhibit invertebrate production or survival of juvenile fish would also reduce or eliminate benthic primary production, but quantitative evidence is lacking. Consequently, a separate section on this topic—effects of deposited sediment on primary producers—is not included in this volume.

EFFECTS ON INVERTEBRATES

Few data are available about the direct effect of suspended sediment concentrations on macroinvertebrates. Possible abrasive action, loss of visual efficiency in feeding, and interference in food gathering by filter-feeding insects (e.g., net-spinning caddisfly larvae) have all been postulated, but only a few reports have produced detailed substantive information.

A few reports have documented the reduction of benthos standing stock or productivity due to turbidity. In their early review, Cordone and Kelley (1961) cited a number of reports on sediment and benthic invertebrates; in several of these, high levels of suspended sediment appeared likely. In all cases, however, the attention of the researcher was obviously on the effect of deposited sediment. Several instances of damage to benthic invertebrates specifically by deposited sediment were cited, but most authors made no mention of the direct effect of suspended sediment.

Investigations in Alaska on suspended sediment concentrations from gold mining operations stand out as the major exception (Wagener and LaPerriere 1985). The authors found that in streams subjected to sediment from placer mining, macroinvertebrate density and biomass decreased significantly in mined streams, relative to unmined reference streams. Although invertebrate changes were first thought to have been the result of settleable solids and mineral fines in bottom substrates (LaPerriere et al. 1983), they were nevertheless correlated with turbidity and suspended solids in a later analysis (Wagener and LaPerriere 1985). However, various heavy metals were associated with the sediments; the relative effects of these contaminants and suspended solids could not be clearly separated.

Another exception appears to be the effect on mollusks, especially unionid clams. Aldridge et al. (1987) reported impaired feeding by three species of unionids in laboratory experiments when sediment was added frequently to simulate suspended solids churned up by intermittent commercial river traffic. Rosenberg and Henschen (1986) suggested sediment particles as the cause of nacre staining in a unionid of the upper Mississippi River.

Brunskill et al. (1973) reported reductions in abundance of filter feeders (some Chironomidae and Trichoptera) as the result of elevated suspended sediment concentrations in a tributary of the Mackenzie River in northern Canada. Forshage and Carter (1974) observed a decrease in invertebrates on multiplate samplers downstream from a gravel-dredging operation that produced suspended sediment in the Brazos River, Texas.

Investigations produced information on two indirect results: (1) increased invertebrate drift, presumably as the consequence of reduced light, and (2) the effect of redeposited suspended sediment at high levels.

Effects on Drift

The effect of light reduction because of turbidity—increased drift—has been well documented. This response has been readily observed in experimental research either in the laboratory or by dosing natural stream riffles with artificially high levels of suspended sediment. Many investigators have reported effects on invertebrate drift related to elevated turbidity caused by suspended sediment (Gammon 1970; Rosenberg and Snow 1975a, 1975b; White and Gammon 1977; Rosenberg and Wiens 1978, 1980; Doeg and Milledge 1991). A number of experiments included dosing a natural stream (or experimental side channels) with suspended sediment; in all cases, the drift of invertebrates increased. This drift response was soon suspected to be the result of a decrease in light reaching the streambed (White and Gammon 1977), similar to the behavioral response in the well-known, night-active diel drift periodicity (Waters 1972; Müller 1974).

The field experiments of Rosenberg and Snow (1975a) in a Mackenzie River tributary resulted in higher drift rates of invertebrates over the short term. However, White and Gammon (1977) speculated that prolonged high concentrations of suspended sediment may reduce benthic populations because of extended drift.

In field observations, Birtwell et al. (1984) observed higher invertebrate drift density, fewer invertebrate taxa, and less food in stomachs of Arctic grayling from silty, mined streams. They concluded that as a result, Arctic grayling were precluded from streams affected by the suspended sediment originating in gold mining operations.

If increased drifting is a response only to decreased light, then it might be expected that populations would fall only to levels that were below carrying capacity (behavioral drift, Waters 1972); consequently, drift would not cause a population reduction to the point of severe depletion. On the other hand, if the drift response was due to abrasion by the suspended sediment particles, or if other physical actions of suspended sediment particles on the animals caused drifting, then elimination might result due to catastrophic drift. Further research on this question would be useful. Where natural levels of suspended sediment have been investigated at levels far below anthropogenic suspended sediment concentrations, no correlation with invertebrate drift was noted (O'Hop and Wallace 1983).

Redeposition of Suspended Sediment

It has been speculated that exceptionally high suspended sediment levels will create accumulations of deposited sediment, either heavy deposits on the streambed or as coatings on stones and other individual substrates. Although little specific research has been applied to this question, an exception is the work of Ciborowski et al. (1979). The authors observed in laboratory experiments that mayfly nymphs entered the drift in response to settled suspended sediment particles (i.e., deposited sediment); the nymphs redistributed to a more favorable habitat without deposited sediment. In this case, the effect of suspended sediment appeared to be indirect, and the observed response was more closely linked to the presence of fine deposited sediment, albeit only a coating, on the nymphs' preferred grazing substrates.

The recent review by Newcombe and MacDonald (1991), which listed most of the papers discussed here, noted mortality as the effect of suspended sediment on macroinvertebrates. They reported lethal results in many publications, but in most cases reductions in standing crop of invertebrates appeared to be due to deposited sediment rather than to suspended sediment; in others, mortality may have been inferred from losses actually due to drift.

On the basis of current knowledge, the direct effect of suspended sediment upon benthic invertebrates does not appear to be a significant influence upon stream invertebrate communities. Prolonged high levels of suspended sediment may, through light reduction, stimulate drift to the point of severe decreases in invertebrate populations. However, in streams with such high suspended sediment levels, it seems probable that heavy accumulations of deposited sediment would also occur, and the deposited sediment might more severely affect benthic populations. In southern Appalachian streams, suspended sediment concentrations of 100,000 mg/L were measured during extreme storms and flood conditions; these concentrations translated to a sediment transport of 9,500 tons/d for each 1 m^3/s of discharge. Such quantities would be thus available as an immense source of deposited sediment (Braatz 1993a).

EFFECTS OF DEPOSITED SEDIMENT ON BENTHIC INVERTEBRATES

The ecology of benthic invertebrates has long been the dominant research interest in stream ecology. Invertebrate biology constituted a major segment of the definitive work on stream ecology by Hynes (1970). However, the importance of the trophic relationship between benthos and fish productivity also has been recognized and researched extensively. Recently, the dependence of fish production upon benthos production has been challenged and modified to include other food resources, for example, in the attempt to resolve the Allen paradox (Allen 1951; Waters 1988). Nevertheless, the reliance of fish upon invertebrate production probably remains substantial in most streams. From the fisheries perspective, as well as others, the effect of excess sediment upon benthic invertebrates has become one of the most important concerns within the sediment pollution problem.

By definition, benthic invertebrates inhabit the stream bottom; therefore, any modification of the streambed by deposited sediment will most likely have a profound effect upon the benthic invertebrate community. Some of the many extensive studies of stream sediment include effects on both the benthos and fish. The influence of sediment deposition on the productivity of benthic organisms as food for fish is one of the most critical problems affecting stream fisheries.

The pervasiveness of deposited fine sediments in the benthic environment, however, was not recognized early. It was sometimes overlooked in stream fisheries studies, in contrast to the much more visible suspended sediment which stimulated the earliest concern about sediment as a stream pollutant. Greater emphasis on the relationship between deposited sediment and benthic invertebrates appeared in later investigations; the relationship now constitutes one of the major subjects in sediment studies.

Most published research dealing with deposited sediment and benthic invertebrates has focused on the general view of dependence by benthos on substrate particle size. In many early fisheries investigations, the principal concern for the invertebrate community—and its relationship to substrate composition—was as a fish-food resource. The review of Cordone and Kelley (1961) stimulated more intensive research into specific effects of anthropogenic sediment inputs.

Although not emphasizing the North American region, the early reviews by Chutter (1969) and later by Ryan (1991) offer many data of comparative significance. Chutter's review was worldwide but emphasized South African streams; he noted that complete smothering of the streambed was necessary to cause large reductions in invertebrates, but small streambed changes could bring about important species changes. Ryan reviewed the effects of both suspended sediment and deposited sediment on plants and invertebrates; he concluded that both of these biotic groups may be reduced rapidly by spates and that full biological recovery required long periods of time.

The relationship between benthos and sediment in streams is incorporated into three major topics: (1) correlation between benthos abundance and substrate particle size, (2) embeddedness of streambed substrates and loss of interstitial space, and (3) change in species composition with change in type of habitat.

First, the positive relationship between benthos abundance and substrate particle size was broadly accepted and appreciated as the main effect of streambed sedimentation. It was the principal influence in fisheries concerns about sediment in the few decades prior to Cordone and Kelley's (1961) review. Second, the dependence by benthos upon interstitial space in the streambed substrate was firmly confirmed by the appreciation of the influence of embeddedness of cobbles on invertebrates. Third, the relationship between substrate particle size and consequent taxonomic changes was appreciated as the result of change in type of habitat.

MAJOR REVIEWS

The most significant publication in its time on stream pollution by sediment was the seminal review by Cordone and Kelley (1961). The review summarized the work on invertebrates reported in about 12 papers published up to that time on sedimentation, including silt generated by mining operations, logging, dam construction, and other sources.

Cordone and Kelley (1961) raised concerns about several sublethal effects on fish, such as lowered body condition, slower growth, and reduced productivity—all correlated with reduced invertebrate food supply; they pointed out the need for continued research on these questions. They concluded that there was " . . . overwhelming evidence that the deposition of sediment in streams . . . has destroyed insect and mussel populations." All reports to that time unanimously agreed that the adverse effect of deposited sediment on benthic invertebrates was serious.

Progress was made in the succeeding decade on the basic ecology of organism–substrate relations by K. W. Cummins (1962, 1964, 1966; Cummins and Lauff 1969). These publications stimulated further concern and research into the relationship between substrate particle size and stream invertebrates. Cummins's contributions were not necessarily concerned with anthropogenic sediment, but they spurred many studies on sediment as a pollutant. If a stream invertebrate biota responded in the way it did in the field studies and experiments of Cummins and his associates, the invertebrate benthos, it was argued, would respond similarly to anthropogenic sediment.

Cummins (1962) reviewed sampling and analysis techniques for both benthic fauna and streambed substrates. He emphasized the effects of the physical characteristics of the stream substrate on the distribution and ecology of benthic invertebrates. In addition to his table of particle-size terminology (Table 1, page 14), Cummins reviewed and cited the literature of technology for sampling and analyzing substrate composition up to about 1960.

The symposium on organism–substrate relations held in 1964 at the University of Pittsburgh's Pymatuning Laboratory of Ecology (Cummins et al. 1966) included the most extensive review of stream ecology at the time (Cummins 1966); 530 references were cited. In this review, Cummins made a strong argument for the inclusion of substrate sampling as standard practice for studies in benthic ecology. He also included a critical discussion of the prevailing technology of substrate sampling and analysis. An increased volume of research on the physical consequences of anthropogenic sediment (i.e., a drastic increase in quantities of small particle sizes) followed in succeeding decades, and the concept of sediment as a pollutant was firmly established.

Up to the mid-1960s, very few papers had dealt with the concept of anthropogenic sediment as a pollutant affecting biotic elements other than fish. Most early papers had considered the effect of turbidity, rather than deposited sediment, as the significant factor. The series of seminars on water pollution at the Robert A. Taft Sanitary Engineering Center, Cincinnati, Ohio (U.S. Public Health Service), in the 1950s was just beginning to consider sediment as a pollutant affecting aquatic life other than photosynthesizing plants and fish (Bartsch 1959; Wilson 1957, 1959). Hynes's (1966) book, the definitive work on aquatic pollution biology at the time, included very little on silt or sediment as a pollutant affecting stream invertebrates.

SUBSTRATE PARTICLE SIZE

A general appreciation of the dependence by benthos on substrate particle size was expressed early: the abundance of benthic invertebrates correlated positively with particle size. Many investigators observed a gradient of benthos abundance across the series of particle sizes (sand-gravel-pebble-cobble), and reported that the greatest abundances occurred in the larger sizes. Over several decades, this principle became well established (Needham 1928; Pennak and Van Gerpen 1947; Sprules 1947; Kimble and Wesche 1975). Minshall (1984) pointed out, however, that the more functional relationship may be between benthos abundance and substrate heterogeneity. Benthos abundance is least in homogeneous sand or silt and in large boulders and bedrock; abundance is greatest in the mixture of heterogeneous gravel, pebbles, and cobbles.

These relationships appear to hold true mainly for the principal taxa of insects available to fish—Ephemeroptera, Plecoptera, and Trichoptera (EPT). For example, higher densities of small burrowing organisms such as chironomid larvae and oligochaetes often occur in silt or muck (Benke et al. 1984); other exceptions include the crustacean *Gammarus* in sand (Waters 1984) and the burrowing mayfly *Hexagenia* in silt (Fremling 1960). Most fish-food organisms—those most available to foraging fish—appear to prosper best in the heterogeneity of pebble and cobble riffles. For many researchers, it followed that additions of anthropogenic sediment would result in loss of the best benthos habitat and

Tailwater fisheries developed below high dams create clear water (and lower temperatures from hypolimnetic releases) through settling of suspended sediment in the reservoir. Such changes also produce alterations in benthic invertebrate taxa, often including tubificid worms coming from the reservoir—which are imitated by the anglers' red "San Juan worm" fly. Salmonids introduced usually include rainbow trout and brown trout. San Juan River, New Mexico.

consequently, reductions in invertebrate populations—an early speculation that has proven overwhelmingly true.

An interesting tangential element in the invertebrate-substrate particle-size relationship was later presented by Culp et al. (1983). Their experiments with different substrate mixtures highlighted the animals' need for organic detritus as food. Lack of organic detritus overrode the effect of substrate: without this food resource, density and biomass of invertebrates decreased significantly regardless of the substrate particle-size mixture.

Almost all early concern about sediment followed the substrate size principle. Siltation was regarded as a simple change from large to small particles, or visually, a covering of the original gravel and cobble substrates with silt and sand. Declines of benthos in heavily sedimented streambeds were the subject of alarms raised by Ellis (1931), who decried the loss of mussels in silted areas as a result of river navigation development in the upper Mississippi River. Since that time, many instances of marked reductions in species and population densities of mussels in the Mississippi River have been noted as a result of human activities, including expansion of navigation and increased river traffic. Generally, these reductions have been attributed to increased sedimentation of previously clean riffles and gravel (Ellis 1936; Marking and Bills 1980). Some authors have documented a wide range of substrate habitat preferences among mollusks (Harman 1972; Huehner 1987). Many reports have documented the loss of species after anthropogenic disturbance, such as dam construction, channelization, increased river traffic, and resulting siltation (Bates 1962; Stein 1972; Harman 1974; Parmalee 1993).

Many reports during the early decades of investigation—prior to Cordone and Kelley's (1961) review—involved the general effect of fine sediment additions that reduced fish food invertebrates. In the Klamath River and tributaries, California, "food organisms" were always observed at lower abundances in streams subject to sedimentation from mining operations (Taft and Shapovalov 1935). Sumner and Smith (1940) reported substantial reductions in "fish food organisms" in tributaries of the American and Yuba rivers, California. These streams had been subject to hydraulic gold mining, a process that produced enormous quantities of sediment, but which is now under much stricter control. Tebo (1955) observed lower abundances of bottom fauna in silted areas of a trout stream in the Coweeta Experimental Forest, North Carolina. Although recognizing the potential of "settleable solids" to adversely affect "fish food production," the Aquatic Life Advisory Committee (ORVWSC 1956) declined to set criteria because of lack of information at the time.

One of the most striking field examples was the 1958 siltation of the Truckee River and a tributary, Cold Creek, in California (Cordone and Pennoyer 1960). Fine sediment from a gravel-washing operation covered stream bottoms up to one foot in depth or more and reduced the substrate to a "hard, bedrock-like" condition. The bottom fauna densities and biomass in both streams were reduced to less than 10% of former levels. The authors concluded that the Truckee River was "severely damaged" up to 10.5 miles downstream from the sediment source.

SUBSTRATE EMBEDDEDNESS

Benthic insects within the EPT group usually inhabit the surface of stones and the interstitial spaces between and beneath large substrate particles such as pebbles and cobbles. Invertebrates in the hyporheic zone—deep within underlying gravel and in contiguous lateral gravel deposits—depend upon the flow of oxygen-containing water through the interstitial spaces. Consequently, research on specific effects of sediment included experiments wherein controlled amounts of sediments were added at increasing levels in order to observe invertebrate behavior and abundance over long periods.

The first such long-term field study on invertebrates was conducted in relation to sediment in mountain streams of central Idaho (Bjornn et al. 1974; Bjornn et al. 1977). Road construction for logging and mining was the main source of fine sediment. The investigations included studies of sediment transport and effects of sediment on juvenile salmonid habitat and benthic insects. The invertebrate research included: (1) investigation of insect abundance and drift with varied sediment composition in a number of natural streams, (2) experimental additions of sediment to riffles in natural streams, and (3) experiments in laboratory streams. The research program extended from 1972 through 1975.

In natural streams, riffles with the most sediment contained the lowest abundance of insects, but small amounts of sediment added to natural riffles did not greatly affect insect abundance or drift.

In the laboratory streams, a major finding was related to the degree of embeddedness of cobbles. The authors concluded that a small amount of fine sediment around cobbles (zero to one-third embeddedness, i.e., the fraction of cobble surface fixed into surrounding sediment) represented the natural condition. At an embeddedness of more than one-third, insect abundance declined greatly (by about 50%), especially for riffle-inhabiting taxa. When a streambed plot was later cleaned of fine sediment, mayflies and stoneflies increased by up to eightfold.

These findings were supported by other experiments where sediments were applied to laboratory stream sections (McClelland and Brusven 1986). The behavior and distribution of riffle insects were observed in relation to the fraction of cobble embeddedness. All species in the EPT preferred the control treatment of cobbles and gravel only; most species responded negatively to increasing amounts of sand and to greater cobble embeddedness. The authors attributed the results to the inability of nonburrowing forms (especially EPT) to gain access to areas beneath embedded cobbles.

Culp et al. (1986) performed an unusual field experiment in Carnation Creek, British Columbia, involving additions of sand (0.5–2.0 mm) to natural stream riffles. They explored the difference in effect of sediment in (1) riffles with low "tractive force" where added fine sediments were deposited and (2) riffles with high tractive force, where the particles, although not suspended, moved along the bottom substrate in a saltating manner. No previous studies had specifically differentiated between these two conditions. Whereas the deposited sediments had little effect (presumably in low amounts or over the short term), the saltating sediments in the high tractive force apparently scoured the animals and caused high catastrophic drift, with a decrease of more than 50% in benthic invertebrate density. Those taxa occupying upper stone surfaces drifted immediately; other taxa with deeper distributions were removed later, after the animals became active in diel patterns. Newbury (1984) provided a discussion of the physical aspects of tractive force acting on sediment particles. Whereas earlier publications speculated on the scouring effect of moving particles upon invertebrates, Culp et al. (1986) appears to be the only one to include empirical data.

Studies in Valley Creek, Minnesota, included the estimate of annual production by hydropsychid caddisflies upstream and downstream from a small reservoir managed as a sediment trap (Mackay and Waters 1986). The reservoir was located a short distance downstream from a major source of chronic sand input (dry gully). Most bottom habitat upstream from the reservoir was sand; the substrate downstream from the dam consisted of cobble riffles. Annual production by three species of hydropsychids was 5.8 g/m^2 (dry weight) upstream from the reservoir and 34.9 g/m^2 downstream from the dam. The downstream increase below the dam was attributed to the feeding behavior of these insects; to construct filtering catchnets, the hydropsychids required solid substrates and interstitial space available in the cobbles downstream from the dam, but lacking in the sandy streambed upstream from the reservoir.

Gammarus has been commonly reported in greatest abundances in fine-particle substrates, particularly sand, rather than in cobble riffles (Pentland 1930; Waters 1965; Marchant and Hynes 1981). This preference has been attributed to avoidance of strong currents (Marchant and Hynes 1981), but *Gammarus* is also known as a strong swimmer and can move easily in loose sand, where it probably feeds on intermixed fine organic detritus as a collector-gatherer. A 5-year study of annual production by *Gammarus* by substrate type in Valley Creek, Minnesota, showed highest production in sand and sand–gravel mixtures and lower production in gravel and cobbles (Waters 1984). Although production decreased in all substrate types as the result of heavy clay deposits, the reduction was most severe in sand that had become embedded with clay; production decreased by 90% in the clay and sand mixture compared to a 65% reduction in cobbles. These results suggested that whereas clean sand may be the preferred substrate for *Gammarus*, it was the clay embeddedness that negatively affected the habitat of *Gammarus*.

An exceptional study on Cullowhee Creek, North Carolina, an Appalachian Mountain stream affected both by logging and organic pollution from pastures, was conducted by Lemly (1982b). The stream was divided for study into three distinct reaches: an upstream, undisturbed uncut control; a reach sedimented by logging operations and residential construction; and a downstream segment additionally polluted with nutrient inputs from horse pastures. Species richness decreased longitudinally through the three reaches: 64 species in the upstream control, 50 species in the reach affected only by sediment, and 36 species in the downstream reach affected by both sediment and organic pollution. Diversity, density, and biomass were all decreased by the sediment alone, but in the organically polluted reach, density and biomass of dipterans were increased. Lemly also suggested that erosion of acidic Appalachian rock and increased input of resulting fine particles reduced the pH of stream water and eliminated some acid-intolerant species. In the sedimented reach alone, a buildup of inorganic particles on the gills and body surfaces of invertebrates was indicated as a source of mortality. Inorganic particles also reduced substrate heterogeneity and interstitial living space, even when the water appeared clear. In the organically polluted lowest reach, the development of *Sphaerotilus* on insect bodies appeared to further increase the adhesion of inorganic particles, thus further reducing gill function. The author suggested a synergistic effect of sediment and nutrient pollution.

Rutherford and Mackay (1986) attributed mortality among caddisfly pupae to silt that was deposited around the pupal case and caused an apparent loss of oxygen supply to the pupa. The authors pointed out that an encased pupa, unlike most other benthic invertebrate stages, cannot move away from poor environmental conditions such as low oxygen levels.

Amphibians were observed to behave similarly to invertebrates when faced with sedimentation. Corn and Bury (1989) observed populations of the tailed frog and three salamanders in 43 small streams in western Oregon. Twenty-three streams were in uncut watersheds; the other 20 had been logged 14–40 years previously. All streams had similar physical factors, except that the streams in logged watersheds had greater sediment deposits and smaller substrate particle sizes. All four species of amphibians had higher density and biomass in the streams in uncut watersheds than in those that were logged. The authors attributed the differences in amphibian populations to the need of larvae to develop in substrate interstices. They emphasized especially the long-term effect of logging along small, low-gradient streams, expressed concern for continued viable populations of these amphibians, and suggested some protective forest management measures.

TAXONOMIC ALTERATION

The long-held general principle of a correlation between invertebrate abundance and substrate particle size was originally formulated on the basis of density of invertebrates. In other words, through the particle-size series of fine to large substrate sizes, numbers of animals generally increased.

There were exceptions, however. For example, clay held very few invertebrates; silt, on the other hand, often held the greatest numbers, especially when the silt was mixed with organic matter. At the other end of the spectrum, boulders and bedrock, instead of holding the most (if the principle held true through the entire size spectrum), often held the fewest invertebrates. It now appears that the principle holds true mainly for insects of the EPT. Part of this discrepancy was probably because insects are more abundant in habitats of greatest heterogeneity, rather than of larger particle size (Minshall 1984). Habitats at both ends of the particle-size spectrum—clay-silt and boulder-bedrock—being essentially homogeneous, support few organisms, and habitats in the center of the spectrum—gravel through cobbles—being much more heterogeneous, hold the most. Another part of the discrepancy—the high abundances in the fine particle sizes—is due to the sometimes superabundance in numbers of small invertebrates of different taxa (e.g., burrowers such as chironomids and oligochaetes). These findings led to the currently accepted principle that when a large-particle habitat of gravel–cobble is changed to silt–sand, a taxonomic alteration also occurs.

In North Carolina, D. R. Lenat and his associates extensively investigated problems of stream pollution resulting from urban runoff, including sedimentation (Duda et al. 1979; Lenat et al. 1979; Lenat 1984; Lenat and Eagleson 1981). Whereas some pollution inputs could be identified from point sources, many could not, owing to the maze of urban development activities and uncontrolled waste disposals. Thus, they concluded that it is a mixture of inputs, including chemical pollution, toxic materials, oil and grease, organic materials, and sediment that degrade urban streams. City paving increases peak flows in streams and thus erosion. Whereas poorer stream conditions were found in large cities due to chemical and organic pollution, streams in smaller towns deteriorated mostly because of sediment and water flow changes (Lenat and Eagleson 1981).

The difficulty in measuring specific pollution sources led these investigators to use a biological monitoring procedure that used abundances of benthic invertebrates to assess stream degradation (Penrose et al. 1980; Lenat 1988). Stream waters of good quality were identified by the greater abundance of pollution-intolerant taxa, such as those in EPT and some Coleoptera; degraded streams contained pollution-tolerant Oligochaeta and some Chironomidae, often in densities higher than for EPT and Coleoptera.

An important value of using macroinvertebrates in an index to water quality lies in the long-term effects that invertebrates will reflect. Life cycles of one year or more will be influenced by changes in hydrological conditions over all seasons; the animals integrate all influences over the longer period. Their populations will reflect synergistic and antagonistic effects of a mixture of pollutants of all kinds— toxic, organic, and sediment (Lenat et al. 1979). Some caution, however, should be used in this regard: the possibility of invertebrate drift (e.g., from a clean upstream section to a polluted reach downstream) might result in incorrect conclusions about the presence of EPT organisms in the polluted reach.

In their major review, Lenat et al. (1979) summarized the use of biological monitoring in streams that receive nonpoint pollution from several sources such as urban runoff, road and highway construction, mining, and agriculture. Sediment was identified as a major component of nonpoint pollution, much more important from all sources (except urban runoff) than chemical pollution.

Lenat et al. (1979) summarized the effects of sediment upon benthic macroinvertebrates.

1. With small amounts of sediment, density and standing stock of the benthos may be decreased due to reduction of interstitial habitat, although structure and species richness may not change.
2. Greater sediment amounts that drastically change substrate type (i.e., from cobble–gravel to sand–silt) will change the number and type of taxa, thus altering community structure and species diversity, but often with increasing densities.

The classic change due to sediment is from a community of EPT to one mainly of oligochaetes and burrowing chironomids. Lenat et al. (1979) supported Cairns's (1977) application of "assimilative capacity" and pointed out that the capacity of a stream to process sediment input without damage to invertebrates depends upon stream gradient and flow. If gradient and flow are sufficient to move away sediments, little change will result. If sediment inputs exceed the stream's capacity to remove them, sediments will deposit and drastically alter the invertebrate community. Lenat et al. (1979) described the difference as being between "habitat reduction" and "habitat change."

The North Carolina investigators further studied the effects of sediment input upon macroinvertebrates in two Piedmont streams, particularly with respect to the above two modes of influence (Lenat et al. 1981). They also described a "stable-sand community" in streambeds where heavy sand deposits at low flow will encourage a periphyton community, which in turn attracts an invertebrate community of grazers. The stable-sand community may be destroyed and washed away under subsequent higher flows. The stable-sand community, however, has the effect of reducing the abundance of those kinds of invertebrates that serve as fish food.

The results of Hall et al. (1984) also demonstrated the difference between taxa that are tolerant and those that are intolerant of fine sediments. In several experiments with artificial channels, the density of chironomids, amphipods, snails, and oligochaetes increased with increasing sediment and embeddedness. Abundances of EPT, when initially present at substantial levels, declined with

increasing sediment. Similarly, Whiting and Clifford (1983) observed fewer species of Ephemeroptera, Trichoptera, Simuliidae, and Amphipoda, and increased
numbers of Tubificidae, in reaches of a stream flowing through Edmonton,
Alberta. They attributed the effects to siltation and other runoff components.

An increase in work on the basic ecology of organism–substrate relationships
confirmed the general conclusion that coarser particles (gravel, pebbles, cobbles)
are preferred by EPT (the most preferred and available fish-food organisms),
whereas fine-particle substrates (sand, silt) are inhabited by chironomid larvae
and other burrowing forms that often are not readily available to foraging fish
(Erman and Erman 1984; Minshall 1984). These are the conclusions most often
reached by investigators studying the effects of sediment from anthropogenic
sources, which almost invariably increase fine particle accumulations and alter
the mix of invertebrate taxa.

SHORT-TERM AND EPISODIC EVENTS

The biological effect of episodic inputs has been found generally to be
temporary. Rapid recovery often results from steep gradients and invertebrate
drift from upstream reaches.

For example, sediment from highway construction at a high elevation in the
Rocky Mountains severely reduced invertebrate density and biomass, but recovery was rapid due to steep stream gradients and quick flushing of deposits (Cline
et al. 1982). In an Ohio stream, sediments from eroding deposits of glacial
lacustrine silt, although natural, simulated episodic events. The glacial silt periodically reduced benthic macroinvertebrates up to 5 km downstream from the
site (DeWalt and Olive 1988). However, after one of the glacial silt deposits was
completely eroded, sediment input ceased, the stream deposits cleared, and drift
from upstream quickly restored benthic populations. In British Columbia, temporary siltation from a pipeline crossing reduced local benthos populations by up
to 74% but benthos recovery was rapid after construction stopped (Tsui and
McCart 1981).

SEDIMENT SOURCES AND INVERTEBRATES

Although sediment from many sources may have much the same effect, some
differences have been often observed. For example, sediment inputs from soil
erosion of agricultural fields will probably be long-term, at least until tillage
practices change. Eroded soil from row-crop fields often includes the finest
particles. Sediment from bridge construction, on the other hand, may be short-
lived. Nonpoint pollution from urban development that contains large amounts
of sediment will most likely contain other pollutants, as will waste from some
mining operations. Consequently, the biological effects of sediment from different
sources may vary in duration, seasonality, and severity, and sometimes with
unknown synergistic effects in a mixture of pollutants.

Agriculture

Whereas sedimentation from agricultural sources is recognized as being
among the most severe, affecting mainly warmwater streams and midwestern

Row crops on the floodplain beside a southern Minnesota trout stream can produce sediment that severely reduces bottom fauna production. The stream, located at left behind trees and shrubs, appears partly protected by a buffering riparian zone.

trout streams, little definitive research on corresponding biological effects has been published. Particularly lacking are long-term studies of the effects of chronic sediment input, such as that which might derive from nonpoint agricultural sources. Unlike work on salmonid reproductive success (discussed later), little precise experimental work has been done in the laboratory on agricultural sediment and invertebrates.

The laboratory stream experiments of Cummins and Lauff (1969) were intended to delineate the basic behavior of benthic invertebrates as related to substrate particle size. Their studies were some of the first experimental research to employ artificial channels to investigate the behavior of stream invertebrates. The authors tested the preference of 10 species of common stream invertebrates for substrate particle size. Animals included mainly EPT, plus one member from each of the genera *Simulium*, *Stenelmis*, *Tipula*, and one snail. Brusven and Prather (1974) conducted similar experiments in laboratory streams, using five species from EPT and Diptera. Although their experiments (like those of Cummins and Lauff) dealt with the basic ecology of the stream invertebrates, they were motivated by concern for sediment as a pollutant resulting from agricultural practice and other anthropogenic sources.

In a specific agriculture-related study, Cooper (1987) reported the effects of heavy sedimentation in Bear Creek, Mississippi, a tributary of the Yazoo River. Sediment input was seasonal; the greatest amounts reached the stream during high rainfall from agricultural fields. The author reported lowest annual production and diversity for invertebrates in those reaches most severely affected by the sediment. During periods of heavy deposition, sensitive species could not survive in some reaches of stream, and invertebrate production was entirely dependent upon pollution-tolerant forms.

Kohlhepp and Hellenthal (1992) estimated annual production by the benthos in Juday Creek, Indiana, which was severely affected by sediment resulting from drainage maintenance (i.e., snag and debris removal). Annual production by five insect species decreased by 78%. The authors emphasized an important functional advantage in using production data, which can be converted into caloric units and thus relate changes in production to stream energy flow through the invertebrate populations.

Using a functional approach, Berkman and Rabeni (1987) studied the effect of siltation in streams located in a largely agricultural area of Missouri. They related siltation specifically to the abundances of various feeding guilds of fishes as they were influenced by the distribution and abundance of invertebrate foods. Effects of siltation differed among the feeding groups; the greatest influence of silt was on the reduction in abundance of the "herbivore" group (central stonerollers) and "benthic insectivores" (several suckers and redhorses, madtoms, and darters), mostly riffle feeding fishes. No feeding guild responded positively to the silt.

Forestry

Studies on effects of logging and other forest management practices constitute a large segment of the literature pertaining to macroinvertebrates. Most of these studies have been conducted in relation to fisheries problems with sediment, usually in areas with steep hillslopes.

Slaney et al. (1977a) were concerned with influences on invertebrate fish food supply as part of the rearing habitat for juvenile steelhead trout. Comparing logged and unlogged areas in interior British Columbia, they reported much lower insect standing stocks and drift from the logged areas. They attributed these effects to the loss of benthic habitat owing to sediment accumulation in the interstitial spaces in stony substrates (i.e., embedded gravel and cobbles). The most severe effects were noted where felling and log skidding were done near the stream. They also reported on experimental additions of sediment, which reduced insect biomass by 75%.

Culp and Davies (1983) also evaluated the effect of buffer strips as part of the Carnation Creek studies in British Columbia. They concluded that without the buffer strips the stream was vulnerable to extensive sedimentation (<0.9 mm particle size) and streambank erosion; also, allochthonous input decreased. The open canopy allowed more sunlight, but primary production did not increase because phosphorus, rather than light, was previously limiting.

Many reports on sediment from forestry sources were concerned with several effects of logging. Other factors affecting macroinvertebrates were elevated water temperatures and increased autochthonous primary production. In some cases, canopy removal and greater sunlight increased production of invertebrates, which followed a greater food supply of autochthonous primary producers (Duncan and Brusven 1985; Noel et al. 1986; Wallace and Gurtz 1986).

Gurtz and Wallace (1984) concluded that the type and size of substrate particles mediated the effect of clear-cutting on streams in the Coweeta Hydrologic Laboratory, North Carolina. They observed that although invertebrates decreased in fine-particle substrates, the benthos increased on moss-covered rock faces due to increased light and primary production by the moss.

Mining

Mining wastes have been blamed for many instances of invertebrate reductions and losses. Some short-term operations have had short-lived effects, but changes in invertebrate populations in streams receiving mining wastes over long periods have proven to be essentially permanent.

Gold-mining operations in western North America have received extensive research attention to deposited sediment and macroinvertebrates. In studies of Arctic grayling streams in interior Alaska, LaPerriere et al. (1983) compared streams in watersheds subjected to placer gold mining with unmined streams. They reported significant reductions in invertebrate populations in the mined streams as compared to unmined ones. Simuliidae (Diptera) and EPT all markedly decreased, whereas Chironomidae increased. These results were attributed to the greater extent of embeddedness of substrates by the mining-derived silt.

Other gold-mining studies included the use of short-term suction dredging, an increasingly popular method but one that entrains aquatic organisms and also causes heavy sediment deposits (Thomas 1985; Harvey 1986; Somer and Hassler 1992). Some suction dredging is recreational, and this does not appear severe with respect to the extent of reductions in benthos downstream. Decreases in macroinvertebrates usually occurred only in the area immediately downstream from the dredging and recovery was rapid. In Thomas's (1985) study, sediment deposits occurred within an area 6–11 m downstream from the dredging site; no effects on benthos were noted farther downstream. Somer and Hassler (1992) reported increased sedimentation and decreased invertebrate populations downstream from dredge sites, but recovery occurred through dredge-hole filling and removal of sediment from the site by subsequent high flows.

Sediment from abandoned mines was an important, long-lasting "disruptive force" in Appalachian streams (Branson and Batch 1972). Benthos abundances were reduced by up to 90%. The studies of Forshage and Carter (1974) on continuous gravel dredging in the Brazos River, Texas, included the report of long-lasting macroinvertebrate declines of up to 97%.

Urban Development

Nonpoint pollution from urban areas often carries large amounts of eroded sediment, but it also carries other substances that affect invertebrates. Road salt, chemical fertilizers, and toxic materials of diverse kinds frequently are included, making it difficult or impossible to differentiate the attributable sources and their effects on the benthos. A common situation is the relatively clean stream emerging from forested or wild land, which then flows through a highly developed urban area to emerge in a heavily sedimented, degraded condition.

A typical example is the study by Hachmöller et al. (1991) on the effect of urbanization in a small suburban stream in Washington State. The stream comprised four distinct types of environment, probably typifying many urban streams: (1) an upper forested site with a rocky, diverse substrate; (2) a channelized reach with an open canopy and a homogeneous bottom substrate; (3) a reach through an urban park with poor water quality; and (4) an urbanized estuarine reach near the mouth of the stream, badly polluted and turbid. The authors' main finding relating to sedimentation was in the difference between the upper, forested reach and the channelized section; species and densities of EPT declined in the homogeneous substrate of the channelized reach.

Roads and Highways

In addition to service roads (usually unpaved) that attend extractive operations such as logging and mining, the construction of public highways also can generate large amounts of fine sediment and severely influence stream invertebrates.

One of the earliest field studies that documented adverse effects of sediment from road construction on stream invertebrates was by King and Ball (1964, 1967). They studied the Red Cedar River, which flowed through the campus of Michigan State University and which received enormous sediment inputs from the construction of Interstate Highway 96 across the southern part of the state. Major concerns were for smallmouth bass habitat, but the authors also noted a " . . . drastically reduced production of the biotic community of the Red Cedar River," amounting to a 68% reduction in periphyton and a 58% reduction in the heterotrophic community.

In Canada's far north, Brunskill et al. (1973), Rosenberg and Snow (1975a, 1975b), and Rosenberg and Wiens (1978, 1980) investigated problems related to oil development in the Mackenzie River region. Experimental additions of sediment to the Harris River, a tributary of the Mackenzie River, resulted in large increases in invertebrate drift which, if sustained, could reduce benthic standing stocks (Rosenberg and Snow 1975a; Rosenberg and Wiens 1978). Rosenberg and Snow (1975b) concluded that deposited sediment was much more detrimental to invertebrates than was suspended sediment and proposed a plan for environmental control of sediment. The plan further recommended development of predictive capability so that sediment inputs could be calculated and consequently controlled at a level that would be naturally flushed out of a stream, thus avoiding long-term effects on the invertebrate community.

In an Ontario highway construction project, invertebrates were denuded from the affected stream area, but populations returned to normal after the construction activity was completed—a rapid recolonization attributed to drift from upstream (Barton 1977). Other highway and road construction projects were studied by Reed (1977) in Virginia and by Chisholm and Downs (1978) in West Virginia. In both cases, benthic invertebrates were reduced substantially by deposited sediment, but drift from upstream recolonized the affected area.

SEDIMENT AND INVERTEBRATE–FISH RELATIONSHIPS

A major concern in the relationship between sediment and invertebrates is the question of the effect on fish production as the result of reduced invertebrate production due to sediment. In a brief review, Hall et al. (1984) pointed out the decided lack of published research on this aspect of the sediment–invertebrate problem. An obvious and important element is the change in macroinvertebrate taxa with sediment—from EPT (common prey of fish) to burrowing forms such as chironomids and oligochaetes (much less available to fish). However, virtually no reports on the effects of these kinds of invertebrate changes due to sediment on fish production exist in the literature. A larger literature exists on the basic relationship between benthos and fish production (Waters 1982, 1993), but research efforts generally have not been concerned with sediment.

A few exceptional studies, however, have included some aspects. Brusven and Rose (1981), using laboratory streams, observed the feeding behavior of a sculpin as affected by various sand–pebble–cobble mixtures. They reported that feeding efficiency appreciably increased with increasing percentages of smaller particles; large amounts of sand apparently made the invertebrates (EPT) more available to the sculpin by reducing insect cover. In other laboratory stream experiments, Crouse et al. (1981) noted decreased production by juvenile chinook salmon as the result of sediment-induced reductions in benthic invertebrates available to the fish.

Murphy et al. (1981), in studies of forest clear-cutting in Cascade Mountain streams in Oregon, reported increases in all components of the stream communities. Higher levels of primary production, increased densities or biomass of benthos, invertebrate drift, salamanders, and trout were attributed in sequence to the increase in light due to canopy removal. They also pointed out that these positive changes may mask the potential negative effects of sedimentation.

Two long-term projects entailing sediment effects included both invertebrates and fish. Alexander and Hansen (1986) added large quantities of sand to a trout stream in Michigan to determine effects on a brook trout population. Pre-treatment studies continued for an initial 5-year period. Experimental additions of sand were made daily for the next 5-year period and posttreatment studies continued for a third 5-year period. After the sand application, benthos densities and biomass dropped to less than half of pretreatment levels and the abundance of brook trout also decreased to less than half of former levels (discussed later). The sand additions affected small, burrowing invertebrates more than larger forms, because of the physical effects of the continually moving sand bed load.

In Valley Creek, Minnesota, a highly productive trout stream, severe sedimentation by clay drastically reduced annual production by both trout and their principal food, *Gammarus pseudolimnaeus* (Waters and Hokenstrom 1980; Waters 1982). The sedimentation occurred in 1970 and 1971, about midway in a 5-year study to estimate production by the *Gammarus*. The sediment apparently was the result of residential development in the upper catchment. Annual production by the *Gammarus* dropped from 271 and 231 kg/ha (dry weight) in the 2 years before sedimentation to 148 and 64 kg/ha—a maximum decrease of about 75%. Some recovery was evident in the next year; production increased to 101 kg/ha. Valley Creek normally contained much sand as well as clean cobble riffles. The sedimentation consisted largely of clay that saturated the streambed; even after the initial turbidity had cleared, the substrate appeared thoroughly saturated with clay. The physical habitat for the fish was not seriously affected, as pools and riffles remained visually clear of sediment, but decreases in trout production (about 60%) were associated with the loss of their main food resource (i.e., *Gammarus*).

CONCLUSIONS

The vitality and health of stream invertebrate populations is tied extremely closely to the particle size of streambed sediments. A change from gravel and cobble riffles to deposits of silt and sand results not only in a precipitate decrease in populations of those invertebrates most important as fish foods, but also a change in species from those inhabiting the interstitial spaces of larger particles to small, burrowing forms less available to foraging fish.

Three specific aspects of the sediment–invertebrate relationship have been described: (1) invertebrate abundance is correlated with substrate particle size; (2) fine sediment reduces the abundance of original populations by reducing interstitial habitat normally available in large-particle substrates (gravel, cobbles); and (3) species type, species richness, and diversity all change as the particle size of the substrate changes from large (gravel, cobbles) to small (sand, silt, clay).

Major interest in the first of these aspects has been directed toward the correlation between invertebrate abundance and substrate particle size, long-held as a general principle. This principle now appears more accurately applied to the Ephemeroptera, Plecoptera, and Trichoptera (EPT) which respond positively to a middle range of particles—gravel to cobbles. Within this range, EPT inhabit interstitial space, which improves with an increase in particle size through the sand-to-cobble range. The EPT group provides the most productive, preferred, and available foods for stream fishes.

Second, the principal effect of excess sediment on the streambed is increased embeddedness of cobbles and other large particles by fine sediments, which fill spaces below and between cobbles, thus reducing habitable area for EPT. The problem of embeddedness was intensively investigated by Bjornn et al. (1974, 1977) and McClelland and Brusven (1980), and thoroughly discussed in the extensive review by Chapman and McLeod (1987). A major advance was the quantification of embeddedness. Bjornn et al. (1974, 1977) measured embeddedness as the fraction of cobble surface fixed in the surrounding sediment and concluded that one-third embeddedness or less is probably the normal condition in streams. Above this level, however, insect populations decline substantially as habitat spaces become smaller and filled. Lenat et al. (1979, 1981) defined this effect as "habitat reduction," which results in a decrease in density and standing stock of existing species.

The third major effect, "habitat change," was also defined by Lenat as the transformation of large-substrate-type deposits to fine sediments. The changed habitat then results not only in the complete elimination of EPT, but also the development of a different invertebrate community—namely, small, burrowing forms such as chironomid larvae and oligochaetes. This change in community invertebrate type often results in greater abundance in numbers, but total standing stocks are often smaller. The burrowing habit of the smaller animals reduces their availability as food for fish. This seems sometimes to have perplexed early investigators who devised the classic correlation between substrate particle size and insect abundance—the correlation held true except for the finest particles of silt or muck, wherein numbers were high. Either major effect—a reduction in EPT or a change to burrowing forms—has a deleterious effect on stream fish populations by reducing the fish food resource.

Construction activities that last only for short times (e.g., road construction, bridges, and pipeline crossings) usually produce transitory effects on invertebrate populations. Whereas the effects of fine-particle deposition may be deleterious to invertebrates in the immediate vicinity of the activity, benthic populations often recover rapidly as the sediment input ceases and deposits wash away downstream. Several factors facilitate rapid recovery of benthic populations:

- insects with flying adults can renew populations quicker than nonflying forms such as mussels and crustaceans;
- undisturbed upstream reaches provide drifting invertebrates for recolonization;
- hydrologic factors (gradient, discharge, current velocity) control the rate of flushing of deposited sediment.

However, in the absence of these factors—especially in slow, low-gradient streams or where continuous erosion occurs on roadways after construction—sediment deposits and reduced or changed invertebrate populations may be long-lasting with respect to both time and downstream distance.

Ongoing operations in agricultural and urban, residential locations present a more long-lasting effect on stream invertebrates, partly because nonpoint sources of sediment are difficult to locate, identify, and control. Logging and other silvicultural operations may have either short-term or long-lasting effects; recovery may be relatively rapid in high-gradient mountain regions, or long-lasting in lower gradients or where poorly constructed logging roads contribute continually eroding sediments.

A contrary effect of logging in small watersheds or along headwater reaches is an increase in macroinvertebrate populations (Murphy and Hall 1981; Murphy et al. 1981; Gurtz and Wallace 1984). Opening the canopy permits more sunlight to reach the previously shaded stream bottom, the additional light increases photosynthesis, and the increased primary production provides more food to invertebrates, particularly grazers. These positive effects may override the deleterious effect of sedimentation in certain conditions. However, such an effect would not be expected in higher-order streams with more naturally open canopies.

Topics that Need Further Work

The specific effect of clay has not been explored sufficiently. Almost all published accounts of the effect of fine sediment on invertebrates have involved silt and sand categories without reference to clay, which may be more invasive into the hyporheic zone.

As opposed to the extensive work on fine sediments and salmonid reproduction, the research on benthic invertebrates includes very little on the effects of particle sizes. Although the paper of Quinn and Hickey (1990) is outside the stated geographical scope of this review, it nevertheless stands as a singular example of extensive research (88 streams in New Zealand) on particle size and benthic invertebrates that is greatly needed. Also related to this topic is the need for greater emphasis on the physical effects of sediment in the hyporheic zone and on invertebrates that inhabit this deeper stratum.

Almost all research published on the effect of fine sediments and benthos involves macroinvertebrates—those organisms, such as most aquatic insects, that are easily visible to the researcher. Virtually no sediment-related work has been done on the effect on the meiofauna, despite increasing evidence of its importance (Benke et al. 1984; Shiozawa 1986). These forms—benthic microcrustaceans such as Cladocera, Copepoda, and Ostracoda, and small insect forms—are crucial as first foods for many fishes, even in small swift streams. The effect of excessive clay deposits on these microinvertebrates in the streambed may be extremely important to stream fisheries and should be investigated.

A greater emphasis on the use of production data, particularly for benthic invertebrates, would greatly improve the analytic value of benthic information in assessing the effect of added sediment (or any other pollutant). Invertebrate annual production, expressed as a rate, would be more relevant to the production rate of carnivorous fish than would the static parameters of abundance and standing stock. A few workers have employed production data (Waters 1984; Mackay and Waters 1986; Cooper 1987; Kohlhepp and Hellenthal 1992), but the ideal practice would be to include production data as a standard technique in invertebrate and sediment investigations.

The paper by Culp et al. (1986) on the effects of high tractive force should spur further work that may be extremely beneficial. Their results suggest a condition wherein continuously moving particles saltating along the bottom might keep benthic populations at depressed levels, even when neither suspended sediment nor deposited sediment may be noticeable.

Further work on location and identification of nonpoint sediment sources is badly needed. This problem affects all components of the stream biological community.

Finally, the recent emphasis on a more holistic approach to stream ecological research and resource management requires that attention be paid to all aspects of surface disturbance in a watershed. Although sediment generation is often an important result of disturbance, sedimentation on the streambed remains as only one of many effects that influence invertebrate production and abundance. Altered flood regimes, changed channel morphology, increased lateral activity, and other hydrologic adjustments all interact with excess sediment (and other pollutants) to influence the invertebrate community.

EFFECTS OF
SEDIMENT ON FISH

The loss or reduction of fish populations has long been associated with turbidity and siltation of streams. Early warnings about the damaging effects of sediment and "erosion silt" (Jordan 1889; Titcomb 1926; Ellis 1936) are representative of this early concern. Specific reasons for fish losses related to sediment have been among the most common objectives of research aimed at identifying causes and remedies.

The greatest attention has been paid to salmonids, particularly juvenile anadromous salmon, steelhead trout, and Arctic grayling in the Pacific Northwest. Additional attention has been paid to sediment problems with Atlantic salmon on the east coasts of Canada and the United States, and inland populations of stream trout (brook, brown, cutthroat). Less concern relative to sediment has been evident for anadromous pink and chum salmon, which do not spend their juvenile years in streams, as do other anadromous species.

Despite early cautions about sediment, however, general conclusions were not at first universally shared. The mining industry, particularly, was unconvinced. For example, from studies in the Rogue River, Oregon, sponsored by the Oregon State Department of Geology and Mineral Industries, Ward (1937–1938) concluded that mining activity was not harmful to fisheries. Ward stated, "Fish live and thrive in rivers carrying large loads of silt," and " . . . neither natural nor artificial erosion up to date has exerted any demonstrable change in the fish food supply. . . . " Upon these conclusions, the *California Mining Journal* for 1938 reported (gleefully, we may presume) that "young fish thrive on mud" (reported by Smith 1940). These conclusions, however, were contradicted by Sumner and Smith (1940), who criticized the experimental methods and analysis leading to Ward's results. After a half-century of the most rigorous research, it is now apparent that fine sediment, originating in a broad array of human activities (including mining), overwhelmingly constitutes one of the major environmental factors—perhaps the principal factor—in the degradation of stream fisheries.

In addition to Ellis's (1936) early paper on "erosion silt," few papers on the subject of sediment were published in that era, although it is apparent that water pollution was beginning to be a subject of concern. Trautman (1933) remarked upon the major causes of several kinds of pollutants that affected fish life, including silt from gravel processing and "soil washings" from newly cultivated fields. Pautske (1938) reported studies on coal washings, including slate and sand particles, which, when experimentally added to habitats of young steelhead and cutthroat trout, caused total mortality. Smith (1940) reviewed a number of studies related to placer gold mining activities in western streams and concluded that (1) though spawning salmon may run upstream through reaches of silty water, they spawn farther upstream in clear water; (2) mining silt is detrimental to incubating eggs in redds; and (3) mining silt blankets stream bottoms and causes high mortalities in the bottom fauna. Shaw and Maga (1943) commented on Ward's (1937–1938) conclusions and conducted their own laboratory experiments with silver (coho) salmon eggs, adding silt derived from placer gold mining in California; they reported much reduced survival of eggs in silted hatchery troughs relative to survival in clear control troughs.

In another early paper by Ellis (1944), which was actually a brief review of work on several aspects of water pollution, he included "suspensoids" as a pollutant and recommended that turbidity in streams should be limited such that light attenuation to one-millionth of incident surface light should not occur at a depth less than 5 m. Ellis (1944) quoted no previous reports that mentioned sediment as a pollutant, but he acknowledged the considerable sediment research at the then-U.S. Bureau of Fisheries laboratory at Columbia, Missouri.

Chapman (1988) credited Harrison (1923) with the first work on sediment and salmonid reproductive success. Harrison's work, however, appeared to be a basic study of the ecology of salmonid egg development related to potential artificial planting of eggs, rather than a consideration of fine sediment as a pollutant.

In subsequent years, concern increased greatly about sediment in salmonid redds, mainly those of western United States and Canadian salmon and steelhead streams. (The work by Wallen [1951] on warmwater fishes was an important exception.) Although field research and laboratory experiments were conducted, most results were presented only in unpublished agency reports; only a few publications were included in primary scientific journals (e.g., reviews by Wickett 1958 on factors affecting pink salmon and chum salmon and by Cordone and Kelley 1961).

Despite the lack of intensive research, it was clear at this early time that fine sediment in salmonid redds was greatly damaging. With loose gravel, survival was high; in redds heavily invaded with fines, survival was low. Cordone and Kelley (1961) also concluded that adult fish can survive suspended silt concentrations usually encountered in natural conditions but that they would succumb when concentrations reach the high levels that damage or coat gill surfaces. They also concluded, on the basis of research done on egg survival, that sedimentation in redds was one of the most important factors that limit natural salmonid reproduction, and that even moderate deposition was detrimental.

The effects of anthropogenic sediments on stream fish are divided into four main categories: (1) direct effect of suspended sediment, including turbidity; (2) effects on salmonid reproductive success in redds; (3) effects on reproductive success of warmwater, or nonsalmonid, fishes; and (4) effects of deposited sediment on the habitat of fry, juvenile, and adult fish.

In many cases, it is difficult to distinguish between the causes of any given effect. The four categories, however, provide a convenient and ecologically meaningful basis for discussion.

EFFECTS OF SUSPENDED SEDIMENT

Suspended sediment produces little or no direct mortality on adult fish at levels observed in natural, relatively unpolluted streams. Early papers were highly speculative, reporting only visual observations of muddy conditions associated with fish kills, or the result of extreme conditions produced in the laboratory (Cordone and Kelley 1961).

Early Observations

In his classic paper, Wallen (1951) reported studies on the effects of suspended sediment on warmwater fishes. His experiments included concentrations up to extremely high levels that caused direct mortalities. Working with 16 species of fish, he found little effect below concentrations of 100,000 ppm montmorillonite clay; fish death occurred at around 200,000 ppm. Mortality was due to clogging of opercular cavities and gill filaments, which presumably impaired respiration. Wallen cited a number of reports of suspended sediment in natural streams much lower than his experimental concentrations, and he thus concluded that suspended sediment was not a lethal condition in concentrations found in nature.

Subsequent results from other researchers on other fishes, particularly salmonids, also suggested deleterious effects at high concentrations. Herbert et al. (1961) reported adult brown trout populations seven times more abundant in a lightly polluted stream than in a Cornish (England) stream heavily polluted with china-clay wastes. They also reported the absence of trout fry in the heavily polluted stream and concluded that there was no reproduction in that stream. Median suspended sediment concentrations were 1,040 mg/L in the lightly polluted river, whereas in the heavily polluted stream the median level was 5,100 mg/L, and short-term maxima were noted up to 100,000 mg/L. Benthic invertebrates were also recorded at much lower densities in the highly polluted stream.

Similarly, Herbert and Merkens (1961) subjected rainbow trout yearlings to different levels of suspended sediment concentration for extended periods of time. They reported substantial direct mortality (>50%) at levels of 270 and 810 mg/L. At lower levels (30 and 90 mg/L) most fish survived up to 185 days. They also observed some disease ("fin rot") and gill damage in fish that died; they emphasized that sublethal concentrations may produce stress in natural river conditions.

Direct Mortality

Intensive investigation of direct mortality due to suspended sediment was not undertaken until recently. In an extensive review, Lloyd (1987) quoted a number of unpublished reports that included results as either fatal or lowered survival; most of these included suspended sediment concentrations from 500 to 6,000 mg/L. Sigler et al. (1984) reported some mortality of very young coho salmon and steelhead trout fry at about 500 to 1,500 mg/L. In laboratory experiments, however, McLeay et al. (1984) reported survival of Arctic grayling underyearlings which had been subjected to prolonged exposure to mining silt in concentrations of 1,000 mg/L. Also, McLeay et al. (1983) reported survival of similar fish acclimated to warm waters in stream cages and subjected to short exposures of concentrations up to 250,000 mg/L.

Obviously, the determination of precise concentrations of suspended sediment that cause acute mortality is difficult and results vary. In field conditions, it may be impossible to distinguish between the effects of suspended sediment and other mortality factors such as heavy metals in mined streams (LaPerriere et al. 1983; Scannell 1988), other toxicants, contaminated suspended sediment particles, and interactions of other effects that by themselves may be sublethal. Servizi and Martens (1991) observed in laboratory tests on acute lethality that smaller fish (coho salmon) were less tolerant to suspended sediment than were larger fish, and that tolerance to suspended sediment was greatest at some optimum temperature (7°C) than at lower (1°C) or higher (18°C) temperatures. In their studies of sediment from placer gold mining and its effects on Arctic grayling, Reynolds et al. (1989) used cage experiments to expose sac fry to suspended sediment in natural streams. Of the sac fry in a mined stream (turbidity >1,000 NTU) 50% died in 96 hours, whereas in an unmined stream only 13% died. Mortality in the mined stream was much higher for sac fry than for fingerlings and juveniles.

Sublethal Effects

Much more information is available on sublethal effects of suspended sediment. Most of these reports were based on laboratory experiments, wherein specific effects were observed critically and often quantified. Effects tested included (1) avoidance and distribution, (2) reduced feeding and growth, (3) respiratory impairment, (4) reduced tolerance to disease and toxicants, and (5) physiological stress (review by Lloyd 1987).

Avoidance and distribution.—Perhaps the most important sublethal effect of suspended sediment is the behavioral avoidance of turbid or silty water, resulting in long reaches or entire streams devoid of fish. Thus, the effect of avoidance may be the total preclusion of resident fish and juvenile anadromous salmonids. This factor destroys a stream as a productive fishery just as surely as if the population were killed. Many published reports have documented such effects.

Avoidance of water muddied from mining silt by spawning adult salmon was observed by Sumner and Smith (1940) in a California river. Avoidance by adult chinook salmon of a spawning stream that contained volcanic ash was reported by Whitman et al. (1982).

DeVore et al. (1980) reported on fish species and biomass in streams in the western Lake Superior region. Some contained red-clay suspended sediment with concentrations up to about 400 mg/L and some were relatively clear. The turbid, warm streams contained many warmwater species, whereas the clear streams contained few species, mostly salmonids. Total fish standing stocks in the turbid streams were about 80–100 kg/ha; in the clear streams (even though having few species) standing stocks were about the same, 50–120 kg/ha.

Birtwell et al. (1984) and Scannell (1988) concluded that Arctic grayling in Alaska streams were confined to clear water only and did not exist in streams heavily exposed to silt from placer gold mining. In work on the Fraser River, British Columbia, Servizi and Martens (1992) observed avoidance by young coho salmon of high suspended sediment derived from gold-mining spoils; they pointed out that coho salmon may move laterally to the sides of a river to avoid high turbidity.

Much research on avoidance of silty water has been conducted in laboratory and field experiments. Juvenile coho salmon (Bisson and Bilby 1982) and young Arctic grayling (Scannell 1988) avoided high concentrations of suspended sediment (as measured by turbidity in NTU). Coho salmon avoided turbidity greater than 70 NTU, and Arctic grayling avoided turbidity greater than 20 NTU. Sigler et al. (1984) also observed that juvenile coho salmon and steelhead trout avoided turbid water in laboratory experiments. McLeay et al. (1984, 1987) observed Arctic grayling that moved in a downstream direction in laboratory streams when subjected to mining silt. Berg and Northcote (1985) observed that juvenile coho salmon exposed to short-term pulses of suspended sediment dispersed from established territories.

In experimental stream channels related to long-term studies on coho salmon in the Clearwater River, Washington, Cederholm and Reid (1987) subjected juvenile coho salmon to three levels of suspended sediment concentrations: clear water (0 mg/L); medium suspended sediment (1,000–4,000 mg/L); and high suspended sediment (4,000–12,000 mg/L). They observed that the fish preferred clear and medium conditions, suggesting that juvenile fish preferentially avoid high suspended sediment conditions in silty streams. Furthermore, they observed evidence of stress in the fish—an increased rate of opercular movement and ''coughing''; sediment accumulations on gill filaments; and declines in prey capture success—at the higher suspended sediment concentrations.

In a different approach involving competition between species, Gradall and Swenson (1982) concluded that red-clay turbidity favored the creek chub over brook trout in sympatric populations in small streams. The creek chub preferred the cover provided by suspended sediment turbidity, whereas brook trout preferred clearer water.

Reduced feeding and growth.—One of the major sublethal effects of high suspended sediment is the loss of visual capability, leading to reduced feeding and depressed growth rate. Several researchers have reported decreased feeding and growth by fish in turbid conditions resulting from suspended sediment. For example, Cleary (1956) and Larimore (1975) noted that turbidity in smallmouth bass streams caused very young fry to be displaced downstream due to the loss of visual orientation. The bass left areas where they fed on the microcrustaceans so important to early fry stages.

Most research on feeding and growth, however, has been experimental. McLeay et al. (1984, 1987) reported impaired feeding ability by Arctic grayling exposed to placer mining silt; Reynolds et al. (1989) reported similar results for Arctic grayling in cage experiments in Alaska streams. Redding et al. (1987) observed little or no feeding by juvenile coho salmon and steelhead trout exposed to suspended sediment in Oregon laboratory experiments, and Berg and Northcote (1985) reported reduced feeding by juvenile coho salmon on drift (brine shrimp) in laboratory tests. In most cases, vision impairment due to suspended sediment turbidity was determined to be the factor that reduced the ability of the fish to capture prey (Sykora et al. 1972; Berg 1982).

Respiratory impairment.—Despite early speculation about gill damage by suspended sediment (Cordone and Kelley 1961; Herbert and Merkens 1961), few reports indicated gill damage and impairment of respiratory function as a source of mortality (McLeay et al. 1987; Redding et al. 1987; Reynolds et al. 1989). Whereas high suspended sediment concentrations may not be immediately fatal, thickening of the gill epithelium may cause some loss of respiratory function (Bell 1973).

Berg and Northcote (1985) reported increased gill-flaring in high turbidities due to suspended sediment; this was viewed as an attempt by fish to cleanse their gill surfaces of suspended sediment particles. Similarly, Servizi and Martens (1992) recorded an eightfold increase in "cough" frequency over controls at suspended sediment concentrations of 230 mg/L. It seems likely that fish have evolved behavioral or physiological adaptations to temporary high concentrations of suspended sediment in order to survive short-term conditions caused by natural spates and floods. Chronic high suspended sediment concentrations that are initiated by anthropogenic sources, however, may not be tolerated.

Studying the effect of Mount St. Helens volcanic ash on chinook and sockeye salmon smolts, Newcomb and Flagg (1983) reported total mortality at very high ash levels (25% ash by volume) but no mortality at less than 5% ash. Based on the appearance of the gills, they suggested that impaired oxygen exchange was the primary cause of death, but they concluded that most airborne ashfalls would not cause acute mortality.

Reduced tolerance to disease and toxicants.—Another potential sublethal effect of suspended sediment is decreased tolerance to disease and toxicants. Several investigators have commented on this possibility, although it does not appear to have been researched intensively. Redding et al. (1987) observed higher mortality in young steelhead trout exposed to a combination of suspended sediment (2.5 g/L) and the bacterial pathogen *Vibrio anguillarum* than in trout exposed to the bacterium alone. Infection in coho salmon fry by a viral kidney disease also resulted in mortality when the fish were exposed to suspended sediment. Goldes et al. (1988) suggested that observed gill lesions were the result of kaolin clay that created a favorable environment for protozoan colonization. McLeay et al. (1984, 1987) reported decreased tolerance of Arctic grayling to an experimental toxicant (pentachlorophenol) in high concentrations of suspended sediment (up to 250 g/L), compared to the toxicant alone.

It appears that suspended sediment induces stress at some raised level of concentration and that such stress tends to reduce the tolerance of fish to a number of environmental factors, including exposure to disease and toxicants. The possibility of further sediment-related problems of disease and toxicants needs more research.

Physiological stress.—Exposure to sublethal levels of suspended sediment may induce physiological stress, which in turn may reduce the ability of the fish to perform vital functions. Redding et al. (1987) reported physiological changes indicative of stress in coho salmon and steelhead trout, including elevated plasma cortisol, plasma glucose, and hematocrits; no direct mortality occurred. The authors concluded that such stress may not be severe, but it may reduce the ability of the fish to feed or resist disease. Servizi and Martens (1992) also reported elevated serum glucose in coho salmon at high suspended sediment levels. In their reviews, Hall (1984a) and Lloyd (1987) included other reports from the unpublished literature that implicated suspended sediment as a factor which induced stress, intolerance, and behavioral problems.

In a recent review, Newcombe and MacDonald (1991) pointed out that in most published studies of suspended sediment and fish, only concentrations were given; they further pointed out that the duration of exposure is essential for more complete understanding of the effects of suspended sediment. They proposed a dose–concentration duration–response model; however, the size of suspended sediment particles, a possible important factor, was not included in the model. The authors also provided a valuable set of three tables listing (1) direct mortality due to suspended sediment, including suspended sediment concentration and duration of exposure (19 papers); (2) sublethal responses (13 papers); and (3) behavioral responses (13 papers). All fish species listed were salmonids. Not all references were for inorganic or suspended sediments, but the high mortality noted in the research cited by Newcombe and MacDonald (1991) points up the need for inclusion of exposure duration. The authors also included a similar table on responses of invertebrates, discussed in a previous section.

Warmwater Fishes

Studies of either direct mortality or sublethal effects of suspended sediment on warmwater fish species are relatively few in the literature. Warmwater streams, although supporting many species, do not appear to attract research support to the same degree as do salmonid streams. Another reason may be that because warmwater streams are often muddy with silt or sand bottoms, their fish species may be perceived to have evolved tolerances to occasional high concentrations of suspended sediment. Some greater effects of deposited sediment have been reported, however, especially on reproductive success, and these are reported in a later section.

The work of Wallen (1951), cited previously, remains the most instructive about suspended sediment effects in warmwater fishes, although not all of the fishes he studied were stream inhabitants. The clogging of gills, and thus respiratory impairment, and induction of disease and parasites have been suggested as effects of suspended sediment (Trautman 1933; Pautske 1938; Wallen 1951).

Despite the observation of seemingly viable fish communities in sometimes extremely turbid and silty conditions, some warmwater species have disappeared over the long term (Larimore and Smith 1963; Smith 1971; Trautman 1981). Muncy et al. (1979), in their extensive review on suspended sediment and warmwater fish, concluded that great variation exists among these species in tolerance to suspended sediment, and that the loss of some species from an otherwise apparently viable fish community eventually may have severe disruptive effects on the system as a whole. They emphasized the need for further research that addresses effects on warmwater species intolerant to suspended sediment.

EFFECTS ON REPRODUCTIVE SUCCESS OF TROUT AND SALMON

The relationship between sediment and salmonid reproductive success has been the subject of greatest concern among all aspects of sediment pollution. Two major reasons for this priority are apparent.

First, salmonids appear as the most favored freshwater recreational fisheries, including adult anadromous salmon on both continental coasts, native anadromous steelhead trout (rainbow trout) in Pacific coast streams and introduced populations in the Great Lakes, as well as inland trout populations (brook, brown, cutthroat, and other trout) across the continent. Second, all North American salmon and trout (except the lake trout) use redds (in flowing waters) in their reproductive strategy, a design that unfortunately functions as a highly efficient "sediment trap," with dramatic and often catastrophic effects on eggs and sac fry.

Through the series of events leading to successful reproduction by salmonids, it is difficult to distinguish between the effects of suspended and deposited sediments, therefore the two components will be treated together, to some extent, in the following discussions. The reason for this special vulnerability is that developing eggs and embryos, as well as newly hatched sac fry, must be supplied by intragravel flow of oxygen-rich water, the chief source of which in many instances is the flowing stream. Deposits of larger particles that may bury the redd completely, or the scouring by floodwaters that eliminates the redd entirely, will have obvious destructive effects.

Three specific effects of sediment on salmonid redds have been recognized: (1) filling of interstitial spaces in the redd by depositing sediment, thus reducing or preventing further flow of water through the redd and the supply of oxygen to the embryos or sac fry; (2) smothering of embryos and sac fry by high concentrations of suspended sediment particles that enter the redd; and (3) entrapment of emerging fry if an armor of consolidated sediments is deposited on the surface of the redd.

In the early period of investigation (pre-1960), these effects were not always recognized separately; often the lack of reproductive success, in its total result, was subjectively associated with a highly visible sedimented stream. Furthermore, the three specific effects cannot always be empirically separated, as indeed they may overlap, in given circumstances. Consequently, in the following discussions they are not treated separately except as the investigator may have handled experiments or specific results.

A fairly large literature has accumulated on methods of measuring conditions in salmonid redds, including substrate particle size and location, distribution of eggs, and presence of invertebrates. Whereas no comprehensive effort was made to include this literature in the present review, some papers most specifically related to salmonid reproduction are included in appropriate sections. The most comprehensive review of methods pertaining to sampling streams and riparian conditions is that by Platts et al. (1983).

Literature Overview

After the stimulus provided by Cordone and Kelley's (1961) review article, research on sediment in salmonid redds increased greatly in the 1960s and 1970s. The most intensive specific effort was soon placed on particle sizes of sediment. The stimulating paper by McNeil and Ahnell (1964) included data on fine sediment in salmon redds in Alaska streams, as well as an emphasis on logging as a major source of fine sediment. The papers by Lantz (1967) and Hall and Lantz (1969) further emphasized logging practices as major sources of sediment; they described the Alsea Watershed Study in Oregon, which included experimental logging in salmonid research.

In the next few decades, many more papers were published on the production of excess sediment from forestry practices and the effect on salmonid reproduction. In addition to the problem of fines reducing water flow and oxygen to redds, sac fry entrapment also stimulated much research in the Alsea study at about the same time (Koski 1966). Other major research programs included long-term studies on the South Fork Salmon River, in Idaho (Platts and Megahan 1975) and on Carnation Creek, British Columbia (Hartman 1982).

Some attention was paid also to Atlantic salmon and inland trout fisheries. These reports included studies on redds of Atlantic salmon in eastern Canada, brook trout in the upper Great Lakes and southern Appalachian regions, and a few other trout populations.

Early Research

In a recent review, Chapman (1988) assigned credit for the first published effort to describe the deleterious effect of fine sediment in salmonid redds to Harrison (1923). At that time, a major effort of fisheries management was the augmentation of fish populations, extension of fish distributions, and restoration of damaged fisheries—all by stocking. Within that management effort, Harrison devised techniques for planting eyed eggs in remote locations, using an "egg planting box" buried in the gravel. He experimented with different sized gravel—from the size of a pea to hickory nut to walnut—variously mixed with clay, silt, and sand. Using these quaint but contemporary units of his day, Harrison then noted survival rates in the different mixtures and reported lower survival in the mixtures with finer sediments.

Further observations of the lack of salmonid spawning and reproductive success in muddy waters were made in the next few decades. The extensive set of observations by Hobbs (1937) of redds made by the introduced chinook salmon and brown trout in streams of South Island, New Zealand, probably stands as the earliest quantitative measures of salmonid egg survival in redds where fine particles (<0.03 in) were also measured. Hobbs concluded that the highest mortality of eggs occurred in the "dirtiest" redds—those with highest percentages of fines. In North America, a few primitive experiments were conducted with gravel and silt, usually in hatchery facilities, slowly building to an accepted conclusion that silty conditions were injurious to developing embryos in salmonid redds (Shapovalov and Berrian 1940; Smith 1940; Shaw and Maga 1943). Wolf (1950), writing in the British *Salmon and Trout Magazine*, reviewed the history of Atlantic salmon fisheries in Canadian tributaries of Lake Ontario. He referred to the immense spawning runs that occurred in the late 1800s but subsequently decreased to almost a total loss. He attributed the decrease to "silting down" of spawning areas by sediment from agricultural cultivation.

In one of the first papers to cite the idea of the salmonid redd as a sediment trap, Moffett (1949) reported on the effects of Shasta Dam (Sacramento River, California), which cleared the river of sediment by deposition in the reservoir and lowered water temperature in a downstream reach. Similarly, Patrick (1976) described the deposition of sediment behind Clark Hill Dam on the Savannah River, which resulted in reduced sediment downstream from the dam and consequently increased numbers of species of algae, arthropods, and fishes.

Research increased slowly in the 1950s. Details of salmonid reproduction, spawning behavior, redd construction, and redd size and density were described for four species of Pacific salmon by Burner (1951). His research was done in connection with attempts to relocate salmon runs in the Columbia River, runs in which spawning fish had been blocked by Grand Coulee Dam. He also noted that fish avoided gravels that were tightly cemented with silt and clay and that almost all successful redds had less than 10% mud, silt, and sand. The classic paper by Stuart (1953) similarly described reproductive behavior of brown trout in Scotland; from laboratory experiments, he concluded that although small, intermittent applications of silt could be washed away by sac fry movements, heavier and continuous silt treatments were lethal.

Further experimental work involved the planting of eggs (Shelton 1955; Gangmark and Bakkala 1960) to investigate if silt prevented sufficient flow of oxygen-containing water. However, Wickett (1954) concluded that "surface silt" on top of the redd did not cause mortality unless it also entered the redd gravel and obstructed water flow. Later, experiments with artificial additions of sediment to spawning gravels supported the conclusion that silt deposited only on top of the redds did not reduce oxygen (Shapley and Bishop 1965).

The recognition of silt and mud as a form of water pollution was expressed early by Ellis (1936), but the paper by Peters (1965) appears among the first work related to salmonid reproductive success to be included in a major publication on water pollution. Peters (1967) also identified agricultural practices that contributed large quantities of sediment to a Montana trout stream. Eyed rainbow trout eggs were placed in Vibert boxes (Anonymous 1951) and stocked in artificial redds located in stream areas variously affected by sediment. Mean suspended sediment concentrations ranged from 20 to 400 mg/L, with overall extremes of 12–1,240 mg/L. The emphasis was on water flow and oxygen supply, both of which decreased at the higher sediment concentrations. Mortality of the rainbow trout eggs increased dramatically with higher concentrations of suspended sediment. This was a rare example of excess sediment from an agricultural source affecting salmonid reproduction.

Early Major Reviews

The review of Cordone and Kelley (1961) was the first comprehensive literature review on the effect of sediment on all components of the biological community in streams, including salmonid reproduction. Although this body of literature was relatively small at the time, their review was a watershed event in historical context; it served as a linch-pin for the immense work on salmonid reproduction to come in following decades. Cordone and Kelley emphasized the importance of fine sediment in salmonid redds and its injurious effect. They cited about 25 publications on this subject, although not all presented sediment data. Their general conclusions were (1) eggs and sac fry are killed as the result of smothering by suspended sediments entering the redd; (2) sediments obstruct the flow of water and its oxygen supply through the redd, causing asphyxiation; (3) continuous applications of small quantities of sediment into the redd are more detrimental than short-term, sudden flushes; and (4) sediment is one of the most important environmental factors that influence the success of salmonid spawning.

Cordone and Kelley (1961) reviewed some of the techniques developed for measuring water flow through the redd. They also identified some sediment sources (e.g., gravel washing and mining effluents). No mention was made of logging practices, which were to become extremely important in later years, especially in the salmon streams of the Pacific Northwest. Very little was cited on the measurement of either particle size of sediments, the identification of "fines" in redds, or the concentration of suspended sediment, all of which later were to receive much more attention.

Many subsequent reviews dealt wholly or in part with the problem of sediment in salmonid redds. Some summaries, dealing generally with the effects of sediment on salmonid productivity, invariably focused on the problem of reproductive success in the redd. These early reviews included

- Phillips (1971)—part of a symposium on forest-land use and streams and a review of most pioneering work in Alaska and Oregon salmon streams;
- Gibbons and Salo (1973)—a large annotated bibliography and discussion, the first major review to indict forestry practice as an important source of sediment adversely affecting salmonid reproduction;
- Platts et al. (1979)—a USEPA report that dealt primarily with sediment particle sizes and their potential effect in the redd;
- Reiser and Bjornn (1979)—part of a larger review of forest management and salmonid stream habitat in the Pacific Northwest.

Most of these early reviews further emphasized forestry practices as sources of sediment in salmonid redds; they provided a great base for the intensive research that followed.

Sediment Particle Size and Gravel Permeability

The effect of suspended sediment, deposited in the redd and potentially reducing water flow and smothering eggs, is a function of sediment particle sizes. Gravel permeability in the redd becomes of first importance, because sediment particle sizes also determine the pore openings in the redd gravel. With small pore openings, more suspended sediments are deposited and water flow is further reduced, compared with larger pore openings.

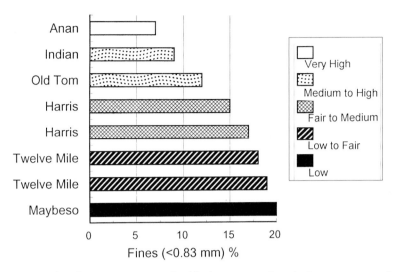

FIGURE 9.—Pink salmon escapement in Alaska streams in relation to percent fines less than 0.83 mm in spawning gravel (from McNeil and Ahnell 1964).

In a brief historical perspective, Chapman (1988) noted some of the pioneering papers in which investigations of fish reproductive success mainly consisted of observations that redds with silt resulted in a lower survival of embryos (e.g., Shaw and Maga 1943; Shelton 1955). With this realization of the importance of fine sediment in spawning gravels, research attention turned to defining the responsible fine particle sizes. Many reports from laboratory and field studies have related survival and emergence to the presence and proportions of fines of specified sizes.

Obviously, such definitions must be based on the function of the particles in their effects on reproductive success. Such an approach was taken by McNeil and Ahnell (1964) in their studies of pink salmon spawning in Alaska. They concluded that the greatest spawning success was in streams with the fewest fines (<0.833 mm, coarse sand and finer) in the redd—apparently the first size definition of "fines" (Figure 9). The selection of the 0.833 mm size was made because they found the best negative correlation between permeability of the gravel bed and the percentage (by volume) of particles less than 0.833 mm (passing through a sieve of that size mesh). The selection of 0.833 mm by McNeil and Ahnell (1964) was further examined by many others.

McNeil and Ahnell (1964) also made one of the first observations of the effect of timber harvest on sediment production and deposition in spawning beds, especially of particles less than 0.833 mm in size. They described a corer-type sampler for collecting material from the spawning bed, including fine sizes, the contents of which were later sieved for grading. The device, which became known as the McNeil corer, was used widely (Figure 10).

Sheridan and McNeil (1968) continued research on the Alaska streams and fine sediments and reported that, although logging increased sediments less than 0.833 mm in spawning gravels, the increase was temporary. The authors sug-

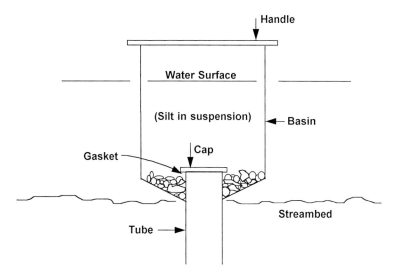

FIGURE 10.—The "McNeil corer," a substrate sampling device retaining fine sediments in spawning gravel studies (from McNeil and Ahnell 1964).

gested that spawning activity by abundant adults assisted in clearing fine sediments from spawning beds. Subsequent investigators also reported this effect (Everest et al. 1987; Bjornn and Reiser 1991). They furthermore suggested that when large particles settle into a redd as the result of redd-digging by female fish, the large particles provide a measure of protection.

Several studies conducted under the auspices of the National Council of the Paper Industry for Air and Stream Improvement (NCASI), New York, contributed valuable insight into the question of the effect of sediment particle sizes. Laboratory experiments on the survival of rainbow trout embryos in different redd-substrate mixtures supported previous results that fines less than 0.8 mm were most damaging, whereas larger particles had much less effect (Hall 1984b). However, Hall stressed the need for more detailed data on the effect of specific sizes of fine sediments.

Another series of NCASI laboratory experiments included three species of Pacific salmon—coho, chinook, and chum—in which survival from eyed eggs to emergence was measured against a series of spawning gravel mixtures (Hall 1986). The range of percentage of fines less than 0.8 mm in the experimental mixtures was 0–50%. Although the control mixture (<0.8 mm = 0) permitted survival of 50–75% for the three species, survival at 10% fines was only 7–10% and in mixtures with more than 10% fines, survival was minimal.

Tagart (1984) also observed an inverse relationship between survival of coho salmon eggs and fines less than 0.85 mm in natural redds in tributaries of the Clearwater River, Washington. In further laboratory experiments with steelhead trout and chinook salmon eggs, Reiser and White (1988) observed little survival beyond 10–20% fines less than 0.84 mm. Eyed steelhead eggs survived better than green eggs, but not beyond the 20% fines level.

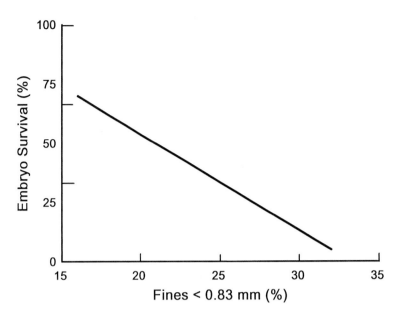

FIGURE 11.—Coho salmon embryo survival (percent) in relation to percent fine particles less than 0.83 mm (by volume) in spawning gravel (from Hall and Lantz 1969).

The level of 20% fines less than 0.8 mm became well established and was accepted by many investigators as the criterion above which significant mortality of embryos could be expected (Figure 11).

Permeability of spawning gravel has received similar attention from researchers investigating salmonid reproductive success. Cooper (1965), investigating sockeye salmon spawning in inland streams of British Columbia, was concerned primarily with water flow through the redd and permeability of redd gravel relative to pore size. With small pore size, more suspended sediments were deposited and water flow was reduced, although Cooper did not define the particle size of fines deposited in redds. Shelton and Pollock (1966) investigated the role of sediment in spawning gravel by measuring the percentage of gravel voids filled by silt. When 35% of voids were filled, salmon egg mortality was 85%; when void siltation was reduced, mortality dropped to as low as 10%. The authors were mainly concerned with oxygen supply. They used an upper part of an artificial channel as a settling basin to control sediment input to the experimental spawning gravel.

Sediment in the Egg Pocket: A Finer Resolution

The use of particle-size categories, particularly as percentages below specific size limits, involved high variability in egg survival, as observed by many researchers. An improvement over particle-size percentage was a measure of central tendency—particularly the geometric mean diameter—calculated from all particles in the spawning gravel smaller than gravel-sized particles (e.g., 25.4 mm).

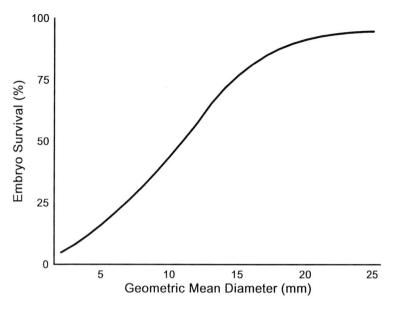

FIGURE 12.—Salmonid embryo survival (percent) in relation to geometric mean diameter of substrate particles in spawning gravel. Survival data are combined survivals of coho and sockeye salmon and steelhead and cutthroat trout (from Shirazi and Seim 1979).

In their studies in the South Fork Salmon River, Platts et al. (1979) questioned the use of percent concentrations of fines and suggested instead the use of the geometric mean diameter of substrate particles, obtaining data from strata in a vertical distribution of redd gravel (Figure 12). The geometric mean, they concluded, was at least as good a predictor of egg survival as percentage fines. They further suggested that a certain quantity of fines is beneficial in the redd, with which Everest et al. (1987) later agreed. Shirazi and Seim (1979) also presented convincing evidence for the superiority of the geometric mean.

Another measure of central tendency, the fredle index, was developed further as another predictor of egg survival (Lotspeich and Everest 1981; Beschta 1982) (Figure 13). The fredle index relates mean particle diameter (such as the geometric mean) to its variance. Lotspeich and Everest (1981) used the data of Phillips et al. (1975) to show a strong correlation between the fredle index and survival to emergence of steelhead trout and coho salmon fry. Sheridan et al. (1984), upon analysis of more than 2,000 streambed samples from pink salmon spawning riffles in southeastern Alaska, reported agreement between the fredle index and percent fines less than 0.83 mm. Young et al. (1991), using cutthroat trout eggs, conducted extensive laboratory experiments with 15 statistics on survival to emergence against substrate composition. They concluded that measures of central tendency (e.g., mean particle size, fredle index), especially the geometric mean, performed better than percent fines in predicting embryo survival.

Reviewing published accounts in a search for criteria that would establish the level of fines causing deleterious effects in salmonid redds, Chapman (1988) noted

FIGURE 13.—Salmonid embryo survival (percent) in relation to the fredle index in spawning gravel (from Lotspeich and Everest 1981).

many laboratory reports relating sediment to egg survival, as well as to the success of fry emergence through overlying substrata. However, he lamented experimental conditions that did not relate necessarily to the immediate environment of the embryos—the egg pocket located in the centrum of the redd. He quoted the use of vertical strata sampling (Platts et al. 1979) as means of acquiring fine-scale data in egg pockets. On the basis of available laboratory data, he supported the hypothesis that permeability and particle size in the egg pocket are greater than in surrounding gravel, i.e., conditions are better in the egg pocket (probably as the result of redd-building activity of spawning females). Chapman made a cogent case for identification of these conditions and, thus, a finer scale of resolution.

Many papers described the characteristics of natural spawning beds, some from the completed redd but a few directly concerned with conditions in an individual egg pocket. The development of the freeze-core technique (Walkotten 1973, 1976; Everest et al. 1980; Lotspeich and Reid 1980; Platts and Penton 1980) made it possible to identify precisely the location of fertilized eggs and the nature and size of gravel and sediment particles in their immediate vicinity (Figure 14). Young et al. (1989) used the freeze-core technique to sample fines in brook trout redds and surrounding gravel; he reported fewer fines in the egg pocket. The frozen cores vary in weight greatly, from about 500 g with the single-probe technique of Walkotten (1973), to the larger ones obtained with the multi-probe unit of Platts and Penton (1986), which requires up to 5,000 lb of lift force. Despite these logistical drawbacks, the frozen cores allow accurate measurements and the

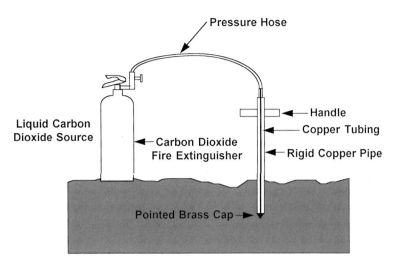

FIGURE 14.—Freeze-core sampler for extracting an intact core of salmonid redd substrate, including sediment from the egg pocket (from Walkotten 1973).

calculation of indices, such as percent fines, geometric mean diameter, and the fredle index, for those particles directly in the egg pocket. Lisle and Eads (1991) recently reviewed the use of several techniques aimed at measuring sediment conditions in spawning gravels, including bulk cores, freeze-cores, infiltration bags, and others.

Chapman (1988) concluded his review with the suggestion that future research should include data on permeability, oxygen, the fredle index, percent of fines less than 0.85 mm and less than 9.5 mm—the two particle sizes for which strong correlations have been found with embryo survival and sac fry emergence, respectively (Tappel and Bjornn 1983). Chapman outlined research phases that include such data from egg pockets. He predicted that when such egg-pocket data are used to correlate with survival, the high variability previously observed would be reduced.

Entrapment of Emerging Sac Fry

The satisfactory completion of the embryo stage for salmonid eggs in the redd is only the first step. Newly hatched sac fry must complete their development in the gravel and upon leaving the protection of their subsurface environment, eventually make their way to the surface of the streambed and to a free-swimming existence. However, sometimes that route is blocked by compacted sediment; entrapment (or entombment) and death of the fry result.

Whereas sediment particle sizes that impede water flow and cause asphyxiation to embryos are small (<0.84 mm), larger particles may permit water flow but still prevent fry emergence through an overhead stratum. First attention was drawn to the emergence problem by Koski (1966), in studies within the Alsea Watershed Study in Oregon. He observed that emergence of coho salmon fry was precluded by sediments less than 3.3 mm. Other work on fry emergence was

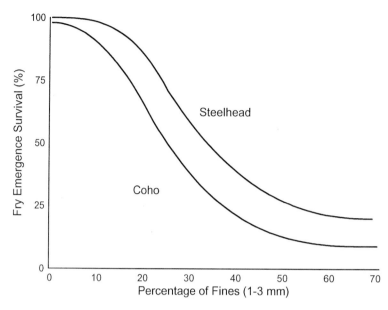

FIGURE 15.—Percent success of fry emerging from spawning redd in relation to percent fines 1–3 mm (from Hall and Lantz 1969).

conducted by Hall and Lantz (1969), Moring (1975a, 1975b), Moring and Lantz (1975), and Phillips et al. (1975). Clearly, past emphasis upon particle sizes and embryo survival in the redd was not the complete view of salmonid reproductive success. Similar results were reported within a fairly narrow range (about 2–6 mm) of particle sizes that prevent emergence (Hall and Lantz 1969; Platts et al. 1979) (Figure 15).

Dill and Northcote (1970) studied the effect of gravel particle size on emergence of coho salmon fry in British Columbia and reported that percent survival was not decreased by gravel sizes (1.9–3.2 cm), but emergence from the gravel by sac fry was delayed. In their recent review, Bjornn and Reiser (1991) concluded that emergence of sac fry from the redd may be impeded by sediments of 2–6.4 mm in percentages above about 10%.

Timber Harvest and Salmonid Reproduction

Timber harvest in the vast coniferous forests of the Pacific Northwest in the United States and in western Canada has received the greatest research attention as an important source of excess sediment in stream salmonid fisheries. The reasons for this attention include, in addition to the immense forest resources, the mountainous topography with steep hillslopes and the great importance of the salmonid fisheries that depend upon appropriate conditions in the many coastal and inland streams. Many watershed features combine to exert tremendous adverse effects by sediment on stream biota, most notably in reduction of reproductive success of all species of salmonids in the area, both anadromous and inland populations (Table 2).

TABLE 2.—Watershed features influencing the risk of sediment limiting salmonid reproductive success (from Everest et al. 1987).

Factor	Low risk	Intermediate risk	High risk
Hydrology	Winter rain hydrograph	Summer rain, winter snow, spring snowmelt hydrograph	Spring-fed hydrograph
Erosion	Surface erosion	Surface and mass erosion	Mass erosion, channel erosion
Geology	Volcanics	Sandstone, siltstones	Granitics
Hillslope	Gentle terrain	Moderate terrain	Steep lands
Stream gradient	High (>5%)	Moderate (1–5%)	Low (<1%)
Stream geometry	Narrow, deep		Shallow, wide
Streamside forest, woody debris	Abundant		Scarce

The several major research programs involving experimental logging in northwestern North America—the Alsea Watershed Study (Oregon), the South Fork Salmon River study (Idaho), and Carnation Creek Watershed project (British Columbia), are all long-term, continuing over several decades. They have contributed greatly to our present knowledge about the effects of sediment in salmonid reproduction. Consequently, this section deals primarily with a review of information having originated in studies of sediment from timber-harvest sources.

In most cases, large-scale commercial timber cutting began before the initiation of sediment research programs, so pre-timber cutting data were not available. The approach, therefore, often was experimental cutting, designed to obtain data from carefully planned logging operations in virgin forests. This approach was the main research design in most of the long-term studies, as well as in some other short-term projects.

Alsea Watershed Study.—The inclusion of the classic paper from the Alsea Watershed Study by Hall and Lantz (1969) in a major symposium on stream trout and salmon (Northcote 1969) was an important advance in the area of forestry–fisheries interactions. In the experimental logging in this Oregon project, three stream watersheds were treated in an experimental forest management program: one clear-cut, one cut in patches and retaining streamside buffer strips, and one uncut as a control (Moring 1975a, 1975b; Moring and Lants 1975) (Figure 3, page 29). Water flow in redds was severely affected in the clear-cut stream, and flow was reduced by more than 75% (Moring 1982). Suspended sediment increased significantly in the clear-cut watershed, but not in the patch-cut or control areas.

Investigators recorded survival of coho salmon eggs and observed a significant inverse relationship between survival and percentage of fine sediment particles less than 0.83 mm. The first report of entombment of emerging fry originated in this program (Koski 1966). Additional experiments on fry emergence resulted in the observation of a significant inverse relationship between successful fry emergence and fine sediment less than 3.3 mm in overhead gravel mixtures.

South Fork Salmon River (SFSR).—Research results from the SFSR study comprised major contributions on the effects of timber harvest on sediment generation and salmonid reproductive success (Platts 1970; Platts and Megahan 1975; Platts et al. 1989). The SFSR project has so far spanned more than a 25-year period. A tributary of the Salmon River in central Idaho (Figure 5, page 32), the SFSR historically was the major producer of steelhead trout and chinook salmon in Idaho. The terrain is steep, with most hillslope gradients of 40–70%—typical of this forested mountain region. A major factor in the SFSR sediment problem is its location in Idaho's central batholith of relatively soft, highly erodible granite. Enormous sediment problems developed early due to logging practices, particularly from roads, in the granitic watersheds.

Before logging began in the SFSR watershed in 1950, annual chinook salmon spawning runs were estimated at about 10,000 (the largest summer chinook salmon run in Idaho) and steelhead trout runs were about 3,000. After logging, runs of chinook salmon were estimated at 1,200 (down by 88%) and steelhead trout runs at 800 (down by 73%). With intense logging in 1950–1965, soil erosion rates increased by 350%, mainly due to road construction. More than 1,000 km of logging roads were constructed, two-thirds of which were on hillslopes of greater than 43% gradient. Forest erosion was accelerated by several storms, after which spawning areas were blanketed by up to 4 ft of sediment deposits (Platts 1970).

After a moratorium on logging and road construction was imposed on the SFSR watershed in 1968, the percentage of fines (<4.7 mm) in spawning gravels decreased from 48 to 25%; the percentage of gravel (4.75–76 mm) increased accordingly (Platts and Megahan 1975; Platts et al. 1989). The authors concluded that recovery under the logging moratorium—which sharply reduced sediment—was accomplished through natural river processes. Apparently, an equilibrium was reached between sediment input from the watershed and the ability of the river to remove sediment, although the low pre-logging sediment concentrations had not been attained.

Carnation Creek Watershed project.—This project also included experimental logging practices (Figure 6, page 34); it contributed greatly to rapidly accumulating knowledge on sediment–fisheries interactions, particularly on reproductive success of coho and chum salmon. Many details, summaries, and historical presentations from the project were included in papers by Hartman (1982), Scrivener and Brownlee (1982, 1989), and Hartman and Scrivener (1990). After logging, coho egg survival to emergence declined from 29 to 16% and chum salmon egg survival declined from 22 to 11%; these reductions were attributed to reduced permeability and oxygen levels in the redd. The size of emerged fry for both coho and chum salmon declined, the number of emerged coho salmon fry dropped to about one-third of previous levels, populations of juvenile steelhead and resident cutthroat trout declined by about 50%, and sculpin populations declined to about one-third of previous levels.

Alaska streams.—Much of the early work on sediment and salmonid reproduction was accomplished in Alaska streams by McNeil and Ahnell (1964) and Sheridan and McNeil (1968). These investigators presented some of the first data on fine sediments related to reproductive success and at the same time first indicted logging practices for erosion and sediment generation to streams (Figure 9, page 90). The work in Alaska stimulated much further research on forest management and timber-harvest sources of sediment.

Northern California.—Harvest techniques in northern California watersheds were studied by Burns (1970, 1972), who reported on effects of various logging practices. Road construction near streams, bulldozer operations on steep slopes, and heavy machine operations directly in stream channels were identified as contributing most in sediment production. Burns also reported substantial decreases in standing stocks of trout and salmon when sediments less than 0.83 mm diameter increased due to these logging practices. Little damage was sustained by fish populations when roads were constructed well away from streams, bulldozer operations were kept out of stream channels, and buffer strips of vegetation were retained along stream edges.

By the early 1970s, concern about the effects of logging on fisheries, including sediment-related problems, and particularly in the Pacific Northwest, led to several major conferences on timber harvest and sediment. In one of the first symposia dedicated to the relationship between forest land use and stream biota (Krygier and Hall 1971), Phillips (1971) summarized the effects of sediment on trout and salmon, listing oxygen reduction in the redd and fry entrapment among other effects. A workshop at the University of Washington resulted in an extensive review and an annotated bibliography (Gibbons and Salo 1973); most of the papers were on salmonid reproduction. Major topics of the workshop addressed sources of sediment in forest management practices and its effects, including the filling of spawning gravel that results in reduction of oxygen supply and entrapment of emerging fry. However, in addition to these problems of excessive sediment, many other aspects of timber-harvest practice were included: effects on water temperature, shading and autochthonous photosynthesis, woody debris in stream channels, and reduction of allochthonous organic matter input.

Recent Major Reviews

Information accumulated by the middle to late 1980s led to several important reviews at a time when much research on salmonid fisheries and forestry-generated sediment had been completed. Almost all of these reviews concentrated on salmonid reproductive success and were primarily related to forestry in the Pacific Northwest.

The Meehan series.—A series of papers under the general title of "Influence of Forest and Rangeland Management on Anadromous Fish Habitat in the Western United States and Canada" was initiated in the late 1970s, edited by W. R. Meehan. The first of these concerned basic habitat requirements (Reiser and Bjornn 1979) and covered (among other subjects) the effect of turbidity on adult spawning, preferred gravel sizes for spawning, and the need for low percentages of fines in redds to allow water and oxygen flow. Other topics included the effect of deposited sediment on invertebrate food production and juvenile fish habitat. Other reports followed subsequently.

NCASI publications.—The research program of the National Council of the Paper Industry for Air and Stream Improvement (NCASI) resulted in the review by Hall (1984a), which emphasized the effects of forestry-generated sediment on salmonid redds and on juvenile rearing habitat. Methods for sampling fine sediment in redds were included, mainly the McNeil sampler (McNeil and Ahnell 1964) and the freeze-core samplers (Walkotten 1973; Everest et al. 1980; Platts and Penton 1980; Lotspeich and Reid 1981). Emphasis was placed on the need for further definition of "fines" in redds, the sizes of fines most damaging to hatching and emergence success, and the relationship of excessive fines to naturally occurring fine sediment concentrations. The need for vertical profiles of redd-gravel composition was pointed out. Many laboratory studies were reviewed, most of which indicated either reduced survival to hatching or reduced emergence by entrapment of emerging fry, or both. However, Hall (1984a) also noted the concern expressed for research on the vertical distribution of sediment in natural redds, in order to define the ambient conditions during successful embryo development and fry emergence. Hall included 67 references in this review, most of which dealt with problems of egg and fry success in salmonid redds.

Forestry–fisheries interactions.—In a book relating forest practices specifically to fisheries, edited by Salo and Cundy (1987), Everest et al. (1987) contributed a major review that summarized the extensive literature on the sediment and salmonid reproduction relationship known up to that time. They termed this relationship a "paradox," drawing the conclusion from the large literature originating in the Pacific Northwest that some fine sediment is natural in salmonid spawning gravels, and in fact may be essential, thus agreeing with Platts et al. (1979). Yet the literature is abundant with many research results that also demonstrate firmly the damaging effects of excessive fine sediment in salmonid redds. The authors pleaded for further work on the "middle ground" between too much and too little sediment (e.g., for a more holistic approach to research on sediment problems).

Everest et al. (1987) also reviewed the voluminous literature on fines in salmonid redds, with some reservation. Laboratory experiments, they pointed out, are often done with synthetic mixtures that probably do not effectively simulate natural redd composition. On the other hand, field studies are often confounded by other effects of forestry practice such as changes in water temperature. They also noted that the measurement of effects directly assignable to sedimentation is more demanding in natural environments and that it is still difficult to produce useful forest management guidelines that relate to sediment. Nevertheless, the evidence from all studies constitutes a strong indictment of excessive fine sediment in salmonid redds; many papers indicate that redd composition including 20% fines of less than 0.83 mm in size is detrimental.

Among biological effects of fine sediments, Everest et al. (1987) listed the blockage of intragravel water and oxygen, direct smothering and suffocation of

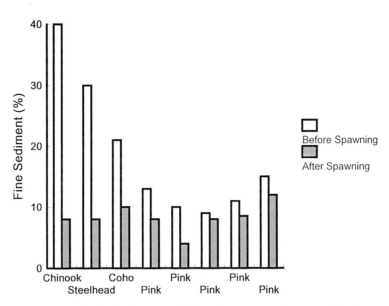

FIGURE 16.—Effect of adult cleaning of fines from spawning gravel during redd construction (from Everest et al. 1987).

eggs and sac fry, and entrapment of emerging fry. They reviewed indirect habitat effects such as the filling of pools and decreases in invertebrate production.

Everest et al. (1987) also pointed out some mitigating factors of fish behavior, mainly that redd-building by female salmonids results in a certain cleaning of fines from the gravel (Figure 16) and that fry behavior such as "coughing" and mucous production on gills assists sloughing of fine particles (Bams 1969). Channel diversity owing to boulders and large organic debris creates "islands" of more suitable spawning habitat, even in a heavily sedimented stream. The point was also made by Everest et al. (1987) that whereas most research included studies of survival in or from redds, little correlative evidence exists on the effect of sediment on populations or the return of adults. Later reports from the Carnation Creek studies, however, now appear to be major exceptions to that observation (Scrivener and Brownlee 1989; Hartman and Scrivener 1990). Everest et al. (1987) included the table of catchment features that influence the risk of limiting populations by damaged spawning habitat (Table 2).

Everest et al. (1987) concluded that all salmonid species can cope with the natural variability in sediments, but their populations can be reduced by persistent sedimentation that exceeds the natural levels under which they have evolved. They closed with a thoughtful and incisive view of future directions in forest management in relation to the salmonid fisheries of the Pacific Northwest. Rather than urge broad guidelines for forest management, however, they suggested that knowledge from a variety of related disciplines be used to tailor operations to specific areas and their changing characteristics.

Chapman's review.—Chapman (1988) presented a critique of previous attempts to define effects of sediment in redds, along with suggestions for the improvement of methods. His review appears to be an outgrowth of a larger report—much broader in scope and more detailed—on the effect of fine sediment that was prepared for the U.S. Environmental Protection Agency (Chapman and McLeod 1987). A historical perspective on research conducted on fines in salmonid redds was included, along with a plea for consideration of conditions in and around the centrum of the egg pocket. The USEPA report also included reviews of the effects of sediment on macroinvertebrate production and juvenile fish rearing habitat.

Forest–range management and salmonids.—The recent reference volume on forest and rangeland management and salmonid fishes, edited by Meehan (1991a) with contributions by many authors, covers a full range of topics within the main subject, including several aspects of sedimentation. This publication constitutes a major development in documenting progress in research on stream salmonid management, although the treatment of sedimentation was a small part. Sediment topics were: sources of sediment (mining and livestock grazing in addition to timber harvesting); specific attention to road construction; juvenile anadromous and resident salmonid rearing habitat; a little on invertebrate production; and mainly, effects on spawning, egg incubation, and fry emergence. The principal chapter on sediment was by Bjornn and Reiser (1991), who contributed an extensive and timely summary of salmonid habitat requirements in streams, including the problem of sediment in redds.

Salmonid Reproduction in Inland Trout Streams

Some of the earliest concerns about sediment affecting salmonid reproduction were expressed in relation to eastern trout streams. Tebo (1957), in a general account of research at the Coweeta Experimental Forest, North Carolina, expressed concern about siltation related to logging practices which he believed would interfere with the success of trout spawning.

Experimentally stocked trout eggs in artificial redds and in Vibert boxes (Anonymous 1951) have failed to survive in silted conditions (Peters 1965, 1967). Furthermore, Harshbarger and Porter (1979) observed better survival of brown trout eggs in natural redds than in Vibert boxes in a southern Appalachian stream. The boxes collected a higher concentration of sediment than did natural redds—an example of an experimental technique that might show a sediment problem when one did not exist.

In the upper reaches of tributaries to Flathead Lake, Montana and British Columbia, Shepard et al. (1985) studied the reproductive success and juvenile densities of bull trout. In experiments with stocked eggs in artificial redds, they reported survival to emergence of 5–50%; survival was significantly correlated with the percentage of fines (30–45%) smaller than 6.4 mm diameter in spawning gravel beds. At 26 stream sites, they found densities of juvenile bull trout ranging from 10 to 1,200 per hectare and correlated with fines less than 6.4 mm diameter in streambeds. Road development in the upper stream watersheds was indicated as the source of sediment. The bull trout population was adfluvial (i.e., spawning in upper stream reaches and maturing in Flathead Lake).

The first studies of sediment sizes and percentages in eastern trout redds were included in the experiments of Cloern (1976), who stocked coho salmon eggs in two Lake Michigan tributaries that were heavily silted. Sediment composition included 17% and 18% fines less than 0.84 mm. Cloern stocked the eggs in Vibert boxes and observed no survival in one stream and only 1.4% in the other.

The southern Appalachian Mountain region contains many trout streams which, although relatively infertile, attract high regard for their recreational fisheries. The brook trout was the only salmonid indigenous to the region; however, introductions of rainbow trout and brown trout have created greater diversity. Although the introduced species, especially the brown trout, have been highly successful in some streams, the reductions of brook trout distribution, especially due to rainbow trout introduction, have caused concern.

Like the Pacific Northwest, hillslopes in the Appalachian Mountains are exceedingly steep. The problem of sedimentation has caused much concern as the result of activities associated with timber harvest, road-building, mining, and other sediment-producing sources. Programs to preserve and protect the remaining brook trout stocks may require special measures to avoid sedimentation. Additional research on the influence of sediment on reproductive success of the brook trout in redds of these southern mountain streams may also be necessary.

West (1979) pointed out the lack of research on characteristics of spawning gravel in southern Appalachian trout streams, as compared with western studies on large salmonids. His studies in trout streams of western North Carolina appear unique in that area. In eight streams, West measured water velocity, dissolved oxygen, permeability, temperature, and bottom composition (particles 0.84 and 3.36 mm diameter) in spawning gravels. Associating these results with published data for other salmonids, he concluded that spawning gravels in the studied streams met requirements for successful embryo development. Except for one stream, the sand component was less than 20% and, with the one exception, the author believed that emergence of trout fry would not be impeded. However, although he reported physical and chemical information, he included no data on fish.

Later, West et al. (1982) experimentally planted eyed brook-trout eggs in two North Carolina streams—one relatively clear and one heavily sedimented—and reported large differences in reproductive success between the two streams. In the clear stream, 87% of the eggs hatched and 11% emerged; in the sedimented stream, 40% hatched and 6.4% emerged. The authors hypothesized that the major mortality factors in the sedimented stream were abrasion of eggs by suspended sediment and entrapment of sac fry by heavy deposited sediment. Trout standing stock in the clear stream was 3.7 times higher than in the sedimented stream. Subsequent observations by Dechant and West (1985) on natural redds of brown trout in two North Carolina streams suggested that sediment deposition during floods accounted partially for decreased reproductive success (along with the effects of scouring and anchor ice).

Additional concerns and studies on the effect of forestry practices on eastern streams were reported by Rutherford (1986) for Atlantic salmon in Nova Scotia, and by England (1987) for brook trout in national forests of Georgia, but details on redd conditions were not included.

Obviously, further study is needed on the effect of sediment in the highly valued salmonid streams of the Appalachian region. The work of West and his associates remains exemplary. In all probability, the southern Appalachian streams, with their steeply sloped watersheds that are highly susceptible to erosion, will show similar vulnerability to sedimentation as those in the Pacific Northwest—most critically with regard to salmonid reproductive success.

Water Source for Redds: Downwelling versus Upwelling

Throughout the extensive research on salmonid reproduction carried out in the Pacific Northwest, the underlying cause of sediment deposition in the redd was considered to be high suspended sediment levels in stream water. In redd construction reported from that region, adult salmonids select locations at the tail of pools or just upstream from riffles, where hydraulic pressure on the streambed increases and causes downwelling into the spawning gravel (Figure 17A).

Here, the source of water for the redd is the stream water that under conditions of high suspended sediment levels delivers sediment into the redd. The suspended sediment content and streambed deposits therefore are critical to sediment composition of redds. Essentially all research has shown a strong correlation between substrate composition (% fines, geometric mean diameter, fredle index) and embryo survival, where stream water was the source of oxygen for the redds (Sheridan 1962).

However, when the water source is groundwater, embryo survival will depend not so much on substrate composition or suspended sediment levels in the stream, but rather on the nature and quality of the groundwater. In this case, successful spawning can occur with redd construction in riffle areas with up-welling water, or even in slow-water locations with fine-particle streambeds.

Considerable evidence from midwestern and eastern regions indicates that the source of water for salmonid redds may be mainly upwelling groundwater (Figure 17B). Benson (1953) observed use of groundwater by spawning brook trout and brown trout in the Pigeon River, Michigan; Latta (1965) correlated young brook trout populations with years of higher groundwater input in the same stream. Webster and Eiriksdottir (1976) observed spawning of brook trout in upwelling water over a period of years in Adirondack streams of New York. They also reported that brook trout detect and select upwelling sites for spawning in laboratory tanks. Fraser (1985) recorded brook trout spawning in shoal areas of an Ontario lake at locations of upwelling groundwater.

Witzel and MacCrimmon (1983b) reported that brook trout selected redd sites containing fine particles in an Ontario stream. In studies of rainbow trout embryo survival in an Ontario tributary of Lake Erie, Sowden and Power (1985) noted survival up to about 50% among 19 natural redds, independent of substrate composition. They concluded that groundwater oxygen content was the limiting factor and suggested that survival above 50% requires a groundwater oxygen content exceeding 8 mg/L.

Similarly, the experiments of MacCrimmon and Gotz (1986) showed better survival of Atlantic salmon embryos—eyed egg to emergence—when water flow in laboratory incubators was upward and sediment was not compacted. Survival

A - Pacific Northwest

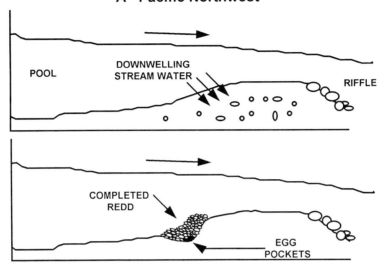

B - The Midwest

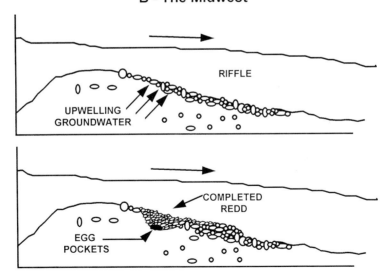

FIGURE 17.—Salmonid redd construction. (A) Downwelling stream water carrying sediment (Pacific Northwest). (B) Upwelling clean groundwater (Midwest). (A is from Phillips 1971.)

Differences in sources of cool water—required in summer for salmonid fisheries—may account for differences in reproductive success. Top: Sandy midwestern streams with cool upwelling groundwater have high reproductive success despite relatively slow currents and sand- or silt-covered streambeds because of sediment-free upwelling water in redds. Crystal Brook, Wisconsin. Bottom: Rocky, steep streams at high mountain altitudes may depend on cool surface runoff for salmonid fisheries where silt-laden stream water, downwelling into salmonid redds, cause mortality to incubating eggs and emerging fry. Bitch Creek, Idaho.

was not related to stream sediment levels; rather, it was attributed to a constant and high rate of upwelling water flow. Such results, of course, may occur only in those natural redds with a source of oxygenated groundwater.

In Lawrence Creek, Wisconsin, a productive but sandy brook trout stream, Hausle and Coble (1976) compared viable eggs and larvae in redds with various percentages of sand, conducted experiments with planted eggs in artificial redds, and observed emergence of embryos in laboratory troughs with various mixtures of sand (<2 mm) and gravel. They found that although hatching and emergence were lower in redd gravels with the higher percentages of sand, successful emergence ran as high as 82% even with the highest sand proportion (25%). These results suggested that brook trout reproduction in sandy midwestern streams may be less susceptible to failure from heavy sand deposition than salmon and trout reproduction in the Pacific Northwest.

In an extensive study of brown trout reproduction in southeastern Minnesota streams, Anderson (1983) concluded that the scouring effect of floods, rather than siltation, was the major factor causing frequent failures of reproduction. The author found evidence of a groundwater source in redds; he suggested that groundwater percolating through the redd would account for reproductive success even when stream waters were silty.

However, not all experiences with eastern species have been similar. For example, experimental studies conducted on Atlantic salmon eggs by Peterson and Metcalfe (1981), in which varying sand proportions were applied to experimental containers, indicated that this species responded with greater survival to emergence with a lower percentage of sand—a response similar to that of western salmonid species. It may be inferred that the differences in reproductive success observed between the two species groups in natural streams is therefore related to differences in redd water source, rather than to differences in species.

Using laboratory incubators with upwelling water, Witzel and MacCrimmon (1981, 1983a) experimented with rainbow trout eggs (homogeneous gravel mixtures) and brook and brown trout eggs (homogeneous gravel and sand-gravel mixtures). Their results were similar to those from western localities and species: survival was low (0–30%) when sand and finer particles constituted 20–30% or more, but higher survival (80–95%) when sand and fines were 0–20%.

Therefore it would be premature to generalize that the difference is due either to species or to continental locales, although the evidence for greater embryo survival in redds with clean groundwater is convincing. Certainly in the trout streams of the upper Midwest (Minnesota, Wisconsin, and Michigan), groundwater sources may well contribute to the great success of both the native brook trout and introduced brown trout, despite the often sandy streambeds of this region.

The question arises: Why is groundwater apparently less available in the West than in the Midwest and East?

One suggestion is that influent streams are common in mountainous terrain and effluent streams are more common in flatter topography. It seems unlikely, however, that influent streams are always the case even in steep-gradient streams. Downwelling water in salmonid redds does not necessarily imply an influent stream. Downwelling in the redd may occur as the result of streambed morphology (i.e., with greater hydraulic pressure just upstream from an elevated riffle). Possibly this type of morphology occurs more commonly in streams with higher gradient.

Such spawning conditions in the Appalachian Mountains, with their steep terrains, may be similar to western conditions: similar sources of stream water and sediment to salmonid redds, with all the attending problems caused by suspended sediment in streams. The work of West et al. (1982) in North Carolina streams suggested that hydrological conditions in streams of the southern Appalachians were more similar to western mountain streams with a stream water supply to salmonid redds.

In streams located at high altitudes (or latitudes) climatic conditions may result in cool runoff in summer, which would favor the salmonid stream with stream water as the source for redds. In the lower altitude of Midwest streams with high runoff temperatures, only those with copious groundwater sources may be cool enough in summer to be suitable for trout or salmon.

The problems of water supply to redds—and salmonid reproductive success relative to sediment—portray an important difference in potential fisheries management between different regions. This question relating to water supply to the salmonid redd is intended as a hypothesis, and it clearly requires further research.

The Holistic Approach to Watershed Management and Salmonid Reproduction

The most recent summaries of the influence of sediment on salmonid reproductive success, particularly those of Everest et al. (1987), Chapman (1988), and Bjornn and Reiser (1991), appeared to signal that the time has come to pause and take stock of salmonid reproduction research for now and proceed further with improved sediment control within the larger scope of salmonid fisheries management.

Some authors have already pointed out the need to proceed beyond the point of ascertaining causes of low egg survival, to determine population effects in the stream, in the ocean stage, and in adult returns. Stowell et al. (1983), using data from the South Fork Salmon River project and other studies, presented a guide for estimating salmonid response from sediment data, including smolt production and adult spawning populations of chinook salmon. Hartman and Scrivener (1990) also made a strong beginning in presenting some preliminary estimates of overwinter survival of fry and juveniles, numbers surviving to the smolt stage and emigration, and marine survival to adult returns. They made a plea for integrating studies of sedimentation (and other forest management impacts) within a broader context of the ecosystem—the holistic view.

In their extensive review, Everest et al. (1987) also urged a holistic approach. As they pointed out, effects of forestry management practices are many and varied. Streams in unforested watersheds also store sediment but still may be favorable for salmon and trout production. The effects of sediment should be studied now in context with other factors, such as water quality and temperature, cover, invertebrate food productivity, large organic debris and other "roughness" elements. Research into sedimentation should expand beyond the study of survival and fry emergence, to assist in developing broader management strategies aimed at maintenance of productive populations, whether resident trout fisheries or adult salmon returns.

Conclusions

The greatest research effort on the relationship between sediment and salmonid reproduction has been applied to anadromous salmon and steelhead trout in the Pacific Northwest and western Canada. In these studies, many researchers have concluded that most reductions in egg-hatching success resulted when fine sediment particles of approximately 0.8 mm and smaller occurred in redds in proportions of about 15–20% or higher (McNeil and Ahnell 1964, and many subsequent authors). Entombment of fry attempting to emerge from redds, another major source of mortality, occurs mainly under high concentrations of larger sediment particles—about 3–6 mm in size—in overlying strata (Hall and Lantz 1969). Subsequent studies suggested that the geometric mean diameter of substrate particles is more precise in predicting hatching success than a simple percentage of fines (Platts et al. 1979). The further development of the fredle index gave additional predictive capability (Lotspeich and Everest 1981; Beschta 1982). The development of freeze-core techniques in sampling redd composition (Walkotten 1973; Lotspeich and Reid 1980) greatly facilitated the analysis of sediment by vertical distribution of sediment in the redd. The freeze-core technique also permitted study of sediments in the redd centrum containing the egg pockets, enabling a "finer scale of resolution" that showed survival closely related to conditions immediately around and among the developing embryos (Chapman 1988).

In the high-gradient streams of the Pacific Northwest, the immediate source of sediment in salmonid redds has been identified as downwelling stream water that carries suspended sediment. Perhaps the same conditions occur in streams of the southern Appalachian Mountains. However, in low-gradient streams of the Midwest and East, upwelling clean groundwater in redds appears to make salmonid spawning more successful than where stream water carrying suspended sediment is the source of water to redds.

With the apparently definitive accomplishments in studies of redd composition and egg-to-fry survival rates, more concern has been expressed recently for the broader view of total watershed management. In addition to production of viable, free-living fry, it is also necessary to discern effects on other life history elements, such as cover and food availability, growth, survival to smolting, and adult return, all of which contribute to attaining the goals of management programs. Such a holistic view has been strongly urged by several authors, including Stowell et al. (1983), Everest et al. (1987), and Hartman and Scrivener (1990).

EFFECTS ON REPRODUCTIVE SUCCESS OF WARMWATER FISHES

The effect of sediment upon reproductive success of warmwater fishes is not well known. Early reports of declines or losses in warmwater populations were broadly correlated with sedimented streams (Trautman 1933; Aitkin 1936; Larimore and Smith 1963; Smith 1971). Sometimes reproductive failures were inferred, but no explicit experimental study has been reported for freshwater stream species.

The major publication on this subject is a comprehensive U.S. Environmental Protection Agency review on the effects of sediment on reproduction and early life of warmwater fishes (Muncy et al. 1979). The majority of papers cited in this report were for lentic species; however, for both lentic and lotic fishes, little explicit research had been done up to the time the review was published. The

· report functioned mostly as a perspective on the problem, and as a detailed plan for the kinds of research that are needed.

Sediment influences upon reproductive success vary with reproductive guilds, that is, groups of fishes within which reproductive behavior is similar, but between which behavior is distinct (Balon 1975). For example, the guilds for which deposited sediment would be expected to be most important include most of the lithophils (dependent upon clean stony substrate) and speleophils (dependent upon interstitial spaces in the stream substrate). Reproductive guilds most tolerant to sediment include the pelagophils (floating eggs) and those lithophils among which the adults clean and guard nests in stony substrates.

Muncy et al. (1979) also reviewed early reported associations between fish populations and sediment. They concluded, however, that reproductive failure due to suspended solids and sedimentation was mainly inferred. Literature that documented specific effects of sediment was scarce or nonexistent.

Reproductive success should be assessed according to all phases of the reproductive cycle. Muncy et al. (1979) considered five reproductive phases: gonad maturation and fecundity, reproductive behavior, embryonic development, larval development, and the juvenile period. The general conclusions on each of these phases were: (1) limited circumstantial evidence exists that sediment limits gonad development; (2) substantial evidence exists that turbidity and sedimentation variously affect time and phase of spawning, which may be negatively correlated with timing of floods and turbidity; (3) egg incubation is particularly susceptible to smothering by sediment; (4) larval stages are more susceptible to suspended sediment than either eggs or adults; and (5) juvenile fish can be directly affected when suspended sediment reduces sight-feeding and respiratory efficiency (discussed in a previous section). Consequently, Muncy et al. (1979) postulated that those species most dependent upon the bottom substrate would be the most susceptible to negative effects of sedimentation.

Muncy et al. (1979) noted that most publications on warmwater fish reproduction were from north-central agricultural regions. Little had been published from western or southern areas.

The authors concluded by identifying several research needs, suggesting three general areas of study: (1) experimental laboratory research on lethal and sublethal effects, in all reproductive phases, at elevated levels of sediment; (2) long-term experimental studies on replicated streams (and ponds) to test modes of sediment action; and (3) use of monitored watersheds to study natural runoff and fish reproduction and to differentiate the effects between sediment and other factors related to agricultural practice, such as pesticide use.

Unfortunately, in the decade and a half since the Muncy et al. (1979) paper, little further research has been accomplished. More study has been applied to marine and estuarine species than to freshwater species, for example, the work of Morgan et al. (1983), who studied the effect of sediment on reproductive success of white perch and striped bass in the Chesapeake and Delaware Canal area.

The only paper specifically on reproductive success of warmwater stream species was that by Berkman and Rabeni (1987), who measured population densities of several fish guilds relative to bottom type in streams of northeastern Missouri. They found that with increased siltation, the abundance of a lithophilous guild with simple spawning (i.e., requiring clean stony or gravel substrate, but with no site preparation or parental care) decreased significantly. These fishes included the golden redhorse, white sucker, central stoneroller, and several other cyprinids. No other guild was affected. Berkman and Rabeni's (1989) paper is an example of the research needed in sediment problems in warmwater streams.

The volume on guidelines for evaluation of human influences on warmwater streams edited by Bryan and Rutherford (1993) contains many individual papers that summarize the effects of anthropogenic sources of pollutants, including sediment. Although biological effects of sedimentation are not included in detail, these papers provide an excellent introduction to stream problems with emphasis on warmwater stream fish habitat. Many problems needing research are identified. Chapter subjects that deal with sedimentation or sediment-related topics include channelization, dredging, sand and gravel extraction, mining, road and bridge construction, timber harvest, and agricultural practice.

In view of the increased concern for fisheries in warmwater streams, especially in the agricultural regions, a heavy commitment to research on sediment and warmwater fish reproduction is obviously called for. Experimental research on the specific phases of reproduction, as outlined by Muncy et al. (1979), would have great potential value in warmwater fisheries management.

EFFECTS OF DEPOSITED SEDIMENT ON FISH HABITAT

Beyond the problems of successful reproduction and sufficient food resources for growth, both of which are susceptible to the deleterious effects of sediment, is the problem of deposited sediment effects on physical habitat—space of adequate quantity and quality to provide for fish needs. These needs include roughness elements on the streambed to provide winter protection for fry against aquatic predators, foraging territories, and sufficient water depths to provide overhead cover for juveniles and adults. Stream features affording these elements constitute rearing habitat; all are subject to severe reductions caused by deposited sediment.

The rearing habitat for juveniles is the most critical. It is well known that the greatest mortality of a given year class or cohort occurs in young stages, and that the strength of a year-class is most often set in some early critical phase (Elliott 1989). Consequently, the sedimentation of juvenile rearing habitat is decisive in its capability to ultimately damage adult fish populations.

Literature Overview

The literature on the subject of deposited sediment and fish habitat has concentrated on rearing habitat. Two major areas of concern have been investigated: (1) winter survival of fry in the interstitial spaces of riffle gravel and cobbles, and (2) depth of pools providing summer cover during growth stages. Like other aspects of this review, the greatest research emphasis within the subject of deposited sediment and fish habitat has been on salmonids.

The overwinter survival of salmonid fry, which requires clean riffles, has been of primary concern because at low water temperatures (<5°C), salmonid fry seek the greater protection of interstitial space in riffles (Chapman and Bjornn

A plunge-pool dam on a small trout stream creates a deep pool for fish habitat. Center notch in the dam assures the sediment-free depth. Forestville Creek, Minnesota.

1969). However, rearing habitat is not as important to those species in which the early fry descend directly to the sea, such as pink salmon and chum salmon.

For most salmonids, rearing habitat remains critically susceptible to deposited sediment. A few investigators have addressed problems with warmwater species, but these have not included the intensity or the detail of experimental research that salmonid research has embodied. Research (and subsequent publication) on physical habitat has not been nearly so extensive as on fish reproduction or invertebrate production but probably ranks third in order of importance.

Early Observations

The early warnings about fisheries in streams affected by sediment applied generally to soil erosion and the consequent silting of streams (Needham 1928; Trautman 1933; Aitkin 1936; Ellis 1936). These investigators did not differentiate between suspended and deposited sediment, their causes, or specific effects. Observations of muddy water or silted streambeds were visually correlated with loss or reduction of fish populations, changes in species from sport to nongame fishes, or decreases in angling success. Often, little effort was made to distinguish among the specific causes of a fishery decline—whether reproductive failure, loss of invertebrate foods, or decrease in physical habitat.

The deleterious effects of bottom accumulations of silt and sand upon fish habitat have been slowly recognized, and in the 1950s some sediment sources were identified. Eschmeyer's (1954) urgent plea to control erosion and sedimentation included mention of southeastern Minnesota's Whitewater River watershed. Severe erosion and excessive sedimentation in the 1930s caused this watershed to lose 90 mi of trout streams (out of an original 150 mi), and the deterioration to "poor condition" of the remaining 60 mi.

Cordone and Penoyer (1960) provided an extensive report on "silt pollution" effects on aquatic life in California's Truckee River and its tributaries. The source of silt was primarily a gravel-washing plant, which caused great turbidity and "striking" silt deposits on stream bottoms. Algae, bottom fauna, and fish were all adversely affected.

The classic paper on this subject was by McCrimmon (1954), whose experimental studies on factors affecting stocked Atlantic salmon fry have stood for many years as landmark research in the ecology of young salmonids. Although McCrimmon included many factors such as temperature and food, the factor that had the greatest deleterious effect upon survival of fry was bottom sedimentation, or deposited sediment. High mortality of the stocked fish occurred as the result of loss of shelter in gravel and riffle spaces, due to filling by sediment and consequent heavy predation by other fishes. Fry survival—particularly winter survival of salmonid fry—in the interstitial spaces of gravel and cobble riffles, so susceptible to loss through sedimentation, remains of major concern.

In their review, Cordone and Kelley (1961) examined the contemporary literature (mostly as informal reports) on deposited sediment. All of those reports suggested serious degradation of available physical habitat by sedimentation. Habitat was lost through filling in of riffle spaces used by small fish, filling of pools occupied by larger fish, and even complete blanketing of stream bottoms with thick deposits. Cordone and Kelley (1961) reviewed early reports such as those by Aitkin (1936) and Trautman (1933), which attributed changes and losses from the midwestern fish fauna generally to erosion and streambed sedimentation. The nine papers quoted began an early distinction between suspended sediment and deposited sediment, and attributed fish mortality to specific effects of deposited sediment. Cordone and Kelley (1961) included in detail the work of McCrimmon (1954) on factors affecting the survival of Atlantic salmon fry.

Fry Survival in Riffles

The need of fry for interstitial space in riffles was emphatically delineated in work on mountain streams in Idaho by Bjornn et al. (1974, 1977). Their research concerned the effect of cobble embeddedness by sediment (<6.35 mm) on the density and distribution of juvenile salmonids (as well as on invertebrates, discussed earlier in this volume). The study involved three parts: (1) experimental additions of sediment to laboratory streams, (2) experimental additions of sediment to a natural stream, and (3) correlative studies of sediment occurrence and fish density among several natural streams. Except for the experimental work of McCrimmon (1954) on stocked fry nearly 20 years previously, this work appeared to be the first to relate experimentally the degree of embeddedness to survival of fry. Their results clearly showed that embedding of cobbles by sediment reduced the physical habitat of juvenile salmonids (mostly chinook salmon and steelhead trout) and consequently the holding capacity of riffle areas for age-0 fish. The decrease in fish density in riffles was greater in winter when fry moved into interstitial spaces, below water temperatures of about 5°C (Chapman and Bjornn 1969; Bjornn 1971). Reductions in fry density were linearly related to the degree of cobble embeddedness. In summer, the greatest effect of excess sedimentation occurred in pools, where decreases in area and depth caused decreases in summer rearing capacity for juveniles.

Similarly, Bustard and Narver (1975) pointed out that salmonid fry overwinter in rubble (cobble) bottoms. In their experiments on British Columbia streams in timber-harvest areas, cutthroat trout showed a strong preference for "clean rubble," as opposed to "silted rubble"; sedimented substrates seriously reduced winter survival of juvenile cutthroat trout. The authors also noted that the problem may be especially important with riffles along stream margins that are dry in summer (and thus may be overlooked in summer surveys). However, the problem may be particularly important to fry in higher winter discharges when they are susceptible to sedimentation effects of timber harvesting in the later season.

Winter survival of fry in sedimented stream bottoms was further studied in Idaho by Hillman et al. (1987), who observed behavior and survival of age-0 chinook salmon in the Red River and its tributaries. They reported that juvenile fish moved away in the fall from heavily sedimented summer habitat in which riffles were 70% embedded and interstitial spaces were filled; experimental additions of cobbles the following year resulted in fivefold increases in winter density of fry. During the following seasons, the added cobbles were removed by scour, sedimented riffles returned, and fry densities fell again in winter to low levels.

Pool Cover as Rearing Habitat

The filling of pools and blanketing of structural cover are two of the most insidious effects of sedimentation. Many streams have been altered to such an extent, and for over so long a time—particularly in midwestern agricultural areas—that little evidence remains with which to make research comparisons. The silted condition may appear to be the pristine condition. Neither does the effect of excess sedimentation appear as simple as the mere covering of the stream substrate. The influence of sediment on channel morphology has been the subject of extensive work in hydrology (e.g., Lisle 1982). Generally, elevated sediment bed loads increase channel width and decrease channel roughness as pools become filled.

Many field studies have related the loss or reduction of fish populations to sedimentation. Saunders and Smith (1965) studied the effect of heavy siltation in a Prince Edward Island brook trout stream; they reported 70% declines in trout populations, in both age-0 and older fish, due to loss of cover by sediment deposits. Brown trout in a Montana stream decreased with increased sedimentation from agricultural sources (Peters 1967). Elwood and Waters (1969) observed drastic declines in a Minnesota brook trout population after catastrophic spring floods that left stream bottoms covered by shifting sand. Scouring apparently also destroyed the eggs or fry of the new year-class. The flood itself did not immediately reduce numbers of the older trout; subsequent decreases that ultimately decimated the trout population were attributed to the total loss of cover caused by the sand deposits.

Other field reports of population declines and species losses included Barton (1977), who observed a decline in total fish standing stock from 24 to 10 kg/ha owing to sedimentation downstream from a bridge construction on an Ontario brook trout stream. However, fish populations returned to former levels after the bridge was completed. In interior Alaska streams, LaPerriere et al. (1983) reported that streams receiving gold mining wastes had degraded spawning and rearing habitats for Arctic grayling. Interstitial gravel spaces were filled, and cement-like substrates were formed. Along with deleterious effects on algae and benthic invertebrates, mined streams were empty of fish, whereas unmined streams in the same area contained healthy populations of Arctic grayling. In British Columbia, Tripp and Poulin (1986) pointed out the deleterious effects of mass soil movements associated with timber-harvest practices resulting in filled stream pools that reduced juvenile fish habitat.

Few papers discuss management attempts to correct problems of habitat loss by deposited sediment. The gradual reduction of the extraordinary sedimentation in the South Fork Salmon River, Idaho, was accomplished mainly by the cessation of logging activity in combination with normal storms and high flows (Platts and Megahan 1975). With short-term operations such as bridge building, sediment deposits tend to be removed quickly and invertebrates and fish are restored, provided that a source of recolonization in nearby populations is present (e.g., Barton 1977).

The most extensive and long-term field experiment relating to sedimentation of fish habitat was that of Alexander and Hansen (1986) in Hunt Creek, a small Michigan trout stream. The experiment included (1) 5 years of pre-treatment study of the trout population (and benthos, previously discussed) in both an upstream 1-mi control section and a lower 1-mi treatment section; (2) 5 years of daily sand addition to the treatment section; and (3) 5 years of post-treatment study. A total of 4,223 yd^3 of sand was added over the 5-year period (stream discharge = 20 ft^3/s). The treatment increased the normal sediment bed load by fourfold, widened and shallowed the stream, and filled pools, whereby the channel became a shallow continuous run of shifting sand with no pools or riffles. Abundance of brook trout declined to less than one-half, the result attributed by the authors to lower survival during summer rearing periods (fry to fingerling) as well as to losses in the egg stage. Not until the fifth year of post-treatment study did fish populations return to normal levels.

Habitat alteration projects often include the installation of structures intended to maintain sediment-free cover, such as undercut banks and plunge pools (described in the Control chapter later in this volume). An example of such a successful program is one in Minnesota, where trout streams had undergone heavy damage from agricultural operations, including sedimentation. Trout populations responded favorably to a habitat improvement program that narrowed channels, deepened pools, and increased riffle areas (Thorn 1988). Increases in trout standing stocks were attributed mainly to increased overwinter survival in the altered reaches.

Restoration of a degraded trout stream. Top: A section of stream in an overgrazed pasture was treated with deflector–bank cover installations and riprap. Bottom: The same section 3 years later with narrowed channel, stabilized stream banks and sediment-free pools for fish habitat. A healthy riparian zone now is protected by fencing to exclude grazing livestock. Middle Fork Whitewater River, Minnesota.

Warmwater Fish Habitat

Sediment problems with warmwater fish habitat appear to be equally severe. King and Ball (1964) speculated that sediment from major highway construction in southern Michigan filled pools or decreased their depths such that smallmouth bass might have been eliminated from some reaches of the Red Cedar River, a once-renowned smallmouth bass stream. Branson and Batch (1972) reported the elimination of some species of fish in a Kentucky stream affected by strip-mining activities and attributed the loss to clay deposits that covered the bottom in places to depths of 2–6 in. A gravel-dredging operation in the Brazos River, Texas, discussed previously in relation to changes in the benthos, was responsible for changes in fish habitat and abundance, notably a decrease in sport fishes (spotted bass, largemouth bass, and bluegill) and increases in other species (Forshage and Carter 1974). Paragamian (1981) concluded from his studies of smallmouth bass in the Maquoketa River, Iowa, that stream-bottom areas of cobble and gravel constituted excellent habitat, with high fish density and standing stock. Small-mouth bass populations were lower in other areas with silt and sand substrates—which constituted poor habitat—although the specific factors responsible were not discussed. Matthews (1984) published a strong indictment of agricultural practices that produced excessive sediment in Wisconsin and destroyed many formerly productive smallmouth bass fisheries in that state's southwestern streams. The comprehensive report by Lyons and Courtney (1990) includes an insightful review of warmwater stream problems in the midwestern United States, with many related references and a compendium of improvement projects in midwestern streams, all of which should stimulate further research on this neglected aspect of deposited sediment.

Major Reviews

Several reviews and major discussions of deposited sediment and fish habitat have been published. Included is the review by Reiser and Bjornn (1979), who addressed many aspects of salmonid habitat. They included a brief discussion of sedimentation in their review of forest and rangeland management effects in western North America. Hall (1984a) reviewed some of the papers published since Cordone and Kelley (1961). Sullivan et al. (1987) included a discussion of the physical aspects of sediment input and transport and resulting effects on the formation of pools, riffles, bars, and other components of stream morphology that form fish habitat.

Everest et al. (1987), in their comprehensive treatment of sediment and salmonid production, pointed out the lack of work on sediment and fish habitat relations (in contrast to the numerous studies of effects on reproductive success). They suggested more research on the relationship between sediment and fishing success, an integrative approach. Finally, Heede and Rinne (1990), in an insightful discussion of hydrodynamic and morphological processes in relation to fisheries management, urged the incorporation of sediment factors into the planning of fish habitat management programs.

Conclusions

The two most important effects of deposited sediment upon the physical habitat of fish are (1) filling of interstitial spaces of riffles, which reduces or eliminates those spaces essential to fry, especially in winter when fry retreat to coarse riffle bottoms for overwinter cover; and (2) reductions of water depth in pools, including the complete loss of pools and cover with heaviest sedimentation, which decrease physical carrying capacity for juvenile and adult fish during summer growth periods.

Most research on both of these effects has emphasized salmonid fishes. Few studies have been reported on warmwater fish species in agricultural areas despite early observations of species losses in streams severely affected by heavy sediment deposits. Fortunately, concern is on the increase for warmwater streams degraded by sedimentation. Lyons and Courtney (1990) suggested that the extension of basic stream ecology, as well as our long-accumulated experience with trout streams, be applied to warmwater problems, an approach that could well result in major advances in warmwater systems management.

CONTROL
OF SEDIMENT

The control of anthropogenic sediment can be viewed in three phases: (1) Prevention—arresting original erosion or obstructing eroded sediment from leaving the site of its origin; (2) Interdiction—capturing and retaining sediment somewhere between the site of origin and the stream; and (3) Restoration—removing sediment from the stream to bring physical conditions back to their original state.

It is greatly preferable to eliminate the source of sediment or the erosive action in the first place. The cost to society increases when intervention occurs further from the source (Reeves et al. 1991), as exemplified by the gradient from prevention to interdiction to restoration. Unfortunately, such "off-site" costs (e.g., loss of fish populations), are viewed as "externalities" in economic analyses and therefore are rarely included in costs to be attributed to the initiating activity.

PREVENTION

Agriculture: Row Crops

A variety of measures are available to prevent erosion on cultivated row crops (Table 3). The problems of soil loss and consequent reduction of productivity have long been recognized by the agriculture industry, and considerable attention has been paid to the development of practices and structures to prevent or reduce such losses. Only in the last decade, however, has greater attention been given to the lost soil as a pollutant that is eroded from cultivated fields and transported off site, often to nearby streams.

Upland erosion control measures.—The agricultural sector has developed technologies highly effective in preventing field soil loss by erosion. In the past, however, these technologies were seldom implemented for stream pollution control, and it appears that the most severe damage to stream fisheries is in the midwestern agricultural region. Further implementation of these technologies needs to be done in this geographic area.

A small volume by Batie (1983), published by The Conservation Foundation, provides a concise but comprehensive review of the soil erosion problem, a brief

TABLE 3.—Agricultural practices for the reduction of erosion from cultivated croplands.[a]

Practice	Method	Purpose	Remarks
Contour planting	Tilling and seeding in rows that follow land contours	Reduce rill erosion, promote infiltration of water	Oldest method of erosion control
Terracing	Grading land to create areas of level or reverse slope	Promote infiltration	Steep, grassed backslopes; most useful on steep slopes
Terrace outlets	Grassed ditches or underground culverts	Provide for removal of excess water from terrace	Size or width based on discharge of 10-year storm
Grassed waterways	Devoting natural drainage routes in fields to grass turf	Prevent erosion and gullying	May be harvested for forage
Strip cropping	Alternating strips of row crops and forage	Reduction of field erosion, promote infiltration	Runoff velocity reduced greatly on grassed strip
Crop rotation	Annual alternation of row crops and forage	Long-term reduction of erosion	Allows time for improvement of soil structure
Spring plowing	Substitution for fall plowing	Avoid bare soil in winter	Reduces wind erosion and water erosion in spring snowmelt
Reduced row crop spacing	Planting rows closer	Reduce erosion of bare soil between rows	Added benefit in increased crop yield
Crop residues	Leaving stubble and other crop remains instead of plowing under	Reduced exposure of disturbed soil, filters sediment in surface runoff	A major element in conservation tillage
Conservation tillage	Avoidance of deep plowing; minimal or no tillage; one-operation tilling and planting	Almost total avoidance of exposed, disturbed soil, increased infiltration	Many variations in shallow plowing; chisel and point plowing; combined with crop residues and reduced irrigation, fertilization, and pesticide use

[a]Based on: USEPA 1973a; NRC 1982; Batie 1983; Christensen and Norris 1983; Clark et al. 1985; Wittmus 1987; Schwab et al. 1993.

history of attempts to control it (starting in the late 1800s), methods of control, and some suggestion of the importance of eroded soil as a pollutant. Other reference books and government publications on erosion control methods include those by Hendrickson (1963), Lum (1977), Goldman et al. (1986), Harlin and Berardi (1987), Wittmuss (1987), and, most recently, Schwab et al. (1993).

The U.S. Soil Conservation Service (USSCS) maintains a number of guides and handbooks that deal with erosion control measures. These include the agency's *Engineering Field Handbook*, which comprises chapters dealing with various

Long-established as means of preventing soil loss from row-crop fields in rolling topography are contour plowing and strip-cropping. Such methods developed by agriculture to save soil also can function to prevent sediment pollution in nearby streams. Southern Minnesota.

structural techniques of soil conservation, such as terraces, diversions, gully treatment, streambank protection, and others. The USSCS's *Field Office Technical Guide* includes a section on standards and specifications for conservation practices. Although these guides and handbooks are intended for USSCS staff use, they may be consulted by the public at USSCS county field offices and other administrative offices.

In the past decade, a number of books and reviews have been published that emphasize environmental concerns for sediment from cultivated croplands as a pollutant (NRC 1982; Overcash and Davidson 1983; Beasley et al. 1984; Clark et al. 1985; Ribaudo 1986).

Some of the earliest methods of erosion prevention included contour planting, (i.e., plowing along level contours across sloping land, rather than straight up and down hillslopes). The furrows from contour plowing retain potential runoff so that water will infiltrate the soil more effectively. Related to this method is the use of terracing, particularly on steeper slopes, which also holds water on the land, or at least slows runoff. In heavy rainstorms, however, terraces may be overwhelmed so that outlets must be provided to carry off excess water. For this purpose, outlets have been designed with the specific intent of preventing scour and gullying, usually with either grassed drainageways or underground culverts (Beasley et al. 1984). Terraces must also include design features that provide for the use of large farm machinery, making this method one of the most expensive.

Practices that involve alternating crops and forms of planting were also implemented in early erosion control. Strip-cropping, crop rotation, and grassed

Grassed waterways through cultivated fields—located on natural waterways in the land surface—trap and filter runoff and help prevent gully formation. Southeastern Minnesota.

waterways were used to slow runoff and reduce its erosive effect by the variety and placement of crops or vegetative cover (USEPA 1973a; Lum 1977). Other practices intended to promote infiltration of water into the soil instead of allowing runoff include plowing in spring instead of in the fall, reducing the distance between crop rows, and planting winter cover (NRC 1982). Trimble and Lund (1982) pointed out that since the 1930s—when soil conservation measures were initiated—annual soil loss rates in the Coon Creek basin, Wisconsin, have dropped to only one-fourth of rates obtained in earlier years.

There have been many kinds of changes to the way soil is broken and turned to prepare it for reception of seed. The traditional moldboard plow that deeply exposes soil, has been partially replaced by alternative methods that come under the overall term of "conservation tillage," which have been developed and used to a limited degree over the past two decades. Contour plowing and strip cropping could be classified as forms of conservation tillage, but more commonly the new methods are chiefly designed to modify the form and depth of plowing, to minimize the extent and duration of soil exposure to water and wind (Batie 1983). Many modifications in plowing have been discussed in the literature but a few of the more common are described below, taken largely from publications by USEPA (1973a), NRC (1982), Batie (1983), Wittmuss (1987), and Schwab et al. (1993).

Conservation tillage.—There are four major conservation tillage practices.

- No tillage (or zero tillage): no seedbed preparation other than slitting or punching openings to receive the seed, leaving crop residue, and with no cultivation during crop production;
- Till-planting: opening furrows, planting seed in the furrows, and closing soil over seed in one operation, with exposure of disturbed soil therefore reduced to near zero;
- Chisel-plowing: loosening soil without inversion, with chisel cultivator or special chiseling plow, also leaving crop residue on the surface, so that chiseling and seeding may be done in one operation;
- Ridge-planting: pushing up ridges of soil into which seeds are planted, leaving old crop residues in the furrow to obstruct runoff and retain sediment.

As indicated above, crop residues from the previous season are often left on the surface (as opposed to being turned over and covered as with traditional deep plowing). The residual vegetation on the surface, like mulch, provides additional protection against erosion and runoff. In some forms (e.g., chisel plowing in wheat stubble), the old residue is hardly disturbed at all. Several studies have evaluated losses of eroded soil among tillage practices, including retention of residues. Compared to traditional plowing, the new conservation tillage practices have resulted in remarkable reductions in losses (Wittmuss 1987).

These new conservation practices appear to be so efficient in preventing erosion on the field that, if widely implemented, this major pollutant affecting warmwater fisheries in midwestern and southeastern regions would be greatly reduced. Unfortunately, conservation tillage does not yet appear to be widely accepted. For example, in the north and central Midwest, conventional plowing decreased only from over 80% to about 70%, and no-till increased by only 1–2%, in the 9 years from 1973 to 1981 (Christensen and Norris 1983). Undoubtedly, additional costs have been an obstructing factor. Long-standing priorities on food production and traditional farm culture may continue to present difficulties in resolving the challenging problem of agricultural sediment pollution in streams.

Floodplain erosion.—In an interesting but provocative paper, Wilkin and Hebel (1982) concluded that a major source of sediment delivered to streams is the cultivated floodplain, from which eroded sediment is transported to streams by floods. From their studies of sediment distribution in an agricultural water-shed—the Middlefork River in northeastern Illinois—they suggested that erosion-control measures applied to the floodplain may be more effective in reducing sediment delivery to streams, compared with control measures in more remote upland sites. Their paper was criticized by Trimble and Knox (1983) who refuted the concept of floodplain soil losses due to floods. They pointed out that flood-plains have been accreting since agriculture was introduced to midwestern wa-tersheds at the time of European settlement, rather than diminishing. Clearly, the perspective of Trimble and Knox was on the question of soil losses or gains on the floodplain; they did not address the question of sediment delivery to streams, nor did they discuss the possible difference in effect on stream pollution between cultivated and untilled treatments on the floodplain (i.e., the stream biological perspective). A more important question to the ecologist or fisheries biologist is what effect cultivation on a given floodplain has upon sedimentation on the streambed immediately downstream from the given floodplain, particularly at a time when floodwaters are receding. Neither Wilkin and Hebel nor Trimble and Knox provided empirical data on that point. Additional questions include the type of conservation tillage (if any) that is employed on the floodplain, whether streambank erosion is controlled, and whether buffer strips between fields and stream are in place. These questions require critical research, most specifically a comparison of sediment delivery to the stream under conditions of floodplain cultivation versus a floodplain in natural vegetation.

Wilkin and Hebel (1982) made several recommendations regarding flood-plain management which appear to be extremely useful as research hypotheses needing study.

- Floodplain lands should be removed from row-crop agriculture and placed in pasture, orchard, or woodland condition.
- Steep lands bordering floodplains should be returned to the forested condition.
- Effective filter strips should be established along upland row-crop fields in order to isolate upland erosion from the floodplains.
- Priority should be placed on controlling erosion from cropped uplands according to their position relative to floodplains and streams.

From a fisheries research perspective, the current situation in cropland erosion control presents an opportunity for needed research. For example, if "pre-treatment" data can be collected now from warmwater streams in areas where conservation tillage has not been practiced but where it is known to be forthcoming, the opportunity to document changes would be invaluable. This procedure would be most efficacious in previously damaged but historically valuable recreational fisheries that have good potential for restoration.

The role of floodplain cultivation in sediment prevention needs immediate attention. The past perspective of soil accretion on floodplains (Trimble and Knox 1983) needs to be turned around and the effort on the stream emphasized. Present research in the literature has stressed agricultural productivity of floodplains—which apparently has increased due to accretion in floods—but research by fisheries biologists and benthological ecologists on effects in the stream is critically needed.

Agriculture: Livestock Grazing

The prevention of sediment generation due to livestock overgrazing lies almost solely in one structural form—fencing of the riparian zone. Whether the stream flows through arid rangelands of the southwest or through the dairy lands of the upper Great Lakes region, the problem may be much the same: livestock are singularly attracted to the riparian zone for drinking water and the more luxuriant forage. However, livestock affect the stream by trampling and destabilizing streambanks, thus widening the channel, decreasing depth, altering current velocity, and causing extensive sediment deposition.

Many of the publications cited in the earlier section "Sources of Sediment" include reports on the use of fencing to prevent livestock grazing in the riparian

Fenced livestock exclosures protect streambanks and sensitive riparian zones, promoting narrower channels, stabilized streambanks, and sediment-free water. The protected area on the right of the photograph is beginning to recover from severe overgrazing. Baker District, Bureau of Land Management, Oregon. (Photograph courtesy of William S. Platts.)

zone, including some long-term experiments. Major reports, including discussion and analysis of fencing as a preventive measure, are by Winegar (1977), Behnke and Raleigh (1979), Chapman (1979), Platts (1981b), several papers in Menke (1983), Armour et al. (1991), and Platts (1991).

Most experiments with fencing have been done in western rangelands but fencing has also been employed along midwestern trout streams to prevent livestock from trampling streambanks. In the latter areas, the intention is not so much to protect the riparian zone as to protect natural streambanks and to keep cattle from the stream itself, while providing for crossings and watering access with floodgates or other means (White and Brynildson 1967).

Fencing experiments in western areas have almost invariably resulted in less sedimentation, better channel and streambank conditions, improved habitat, and higher densities or standing stocks of benthos and fish (Behnke and Raleigh 1979; Dahlem 1979; Keller et al. 1979; Van Velson 1979; Duff 1983; Marcuson 1983; Stuber 1985).

Fencing also can be used as a means of restoring a stream fishery previously damaged (Claire and Storch 1983; Platts 1981a). In addition to preventing livestock from trampling streambanks, fencing also protects the rich vegetation of a riparian zone, which may act to filter sediment in runoff from grazed, as well as cultivated, uplands (Elmore 1992).

Although fencing appears to be the principal structural method to protect riparian zones, several authors have pointed out that fencing should be used only as part of an overall management plan. In an incisive summary to a forum on grazing and riparian-stream ecosystems, held in Denver, Colorado, in 1978 (Cope 1979), Chapman (1979) outlined the overall problem and pointed toward some general solutions. He concluded that of the 150 million acres of public grazing lands, 83% were in a degraded condition, and most of the 19,000 mi of streams included within the area were also degraded. However, he emphasized that 19,000 mi of fencing was impracticable and urged alternatives, such as increased and improved use of uplands for grazing by livestock.

A number of suggestions have been made regarding grazing strategies that may not require continuous fencing to preclude livestock from riparian zones or streams. The general principle that has emerged asserts that livestock grazing should be managed separately in riparian and upland areas, using fencing as part of the general strategy but not the only one (Berry 1979; Behnke and Raleigh 1979). Platts and 13 coauthors (1983) urged that the riparian zone be viewed as a separate ecosystem, i.e., its study and management should be integrated into an overall, multiple-use concept of rangeland management. They also emphasized the need for site-specific research on riparian-stream fishery interactions. The point is that conservative grazing in the riparian zone can benefit both the stream and the land user.

The recent volume edited by Naiman (1992) treats the watershed in a holistic sense but also addresses specifically the influence of livestock grazing on the riparian zone. General alternatives in grazing management include some system of resting pasture units (including the riparian zone) and rotating stock among pasture units (Behnke and Raleigh 1979; Platts 1981b; Platts and Nelson 1985a; Johnson 1992). Platts (1991) outlined a large variety of such options, with different alternatives for sheep and cattle, but he concluded that attempts to evaluate different strategies remain preliminary.

Specific recommendations by several authors included giving complete protection to selected fisheries of high value and sensitive areas such as springs; developing watering areas away from streams and springs; and breeding for cattle less attracted to riparian areas (Behnke and Raleigh (1979). All agreed that intensive research in a multidisciplinary approach is needed to address the problem of integrating stream and fishery values with the need for providing livestock forage.

Forestry

Control measures to prevent erosion from logging roads remain a paramount concern. There is much more literature on sediment production from logging roads and the preventive measures used to reduce erosion from roadways and associated surfaces than the relatively moderate amount selected for the present review. This literature has appeared recently, mostly over the past two decades. It has been generated chiefly in the Pacific Northwest of the United States and western Canada, and the southern Appalachian Mountains—all regions with steep hillslopes (Table 4).

TABLE 4.—Methods for the reduction of erosion from logging roads.[a]

Design feature	Method	Purpose	Remarks
Road placement	Avoid roads near streams, on steep slopes, or in inner valley gorge	Reduce erosion, mass soil movement, and transport to streams	Placement often single most important factor
Road length	Few roads and as short as possible	Reduce total area of exposed roadbed	Length greatly reduced by skyline and helicopter yarding
Road width	Keep narrow, but accommodate equipment	Reduce excavation, less fill, less area of exposed roadbed	Less disturbance reduces probability of mass soil failure
Road grade	5–15%, not flat, minimum 3% for drainage	Avoid rapid runoff on roads	Switchbacks and sharp turns require culverts or other measures
Road surface	Surface roadbed with gravel, crushed rock, oil, or salt	Reduce direct erosion from roadbed	Caution: oil and chemicals may pollute streams
Cut slopes	Vertical or near vertical cut	Reduce excavation and erosion of slope	Vertical cuts may promote mass soil movement in unconsolidated material
Fill slopes	Avoid road drainage and woody debris in fill	Stabilize fill slope	High probability of mass failure
Road drainage	Outslope drainage on low grades; inside drainage on steep grades	Disperse drainage to reduce gullying	Inslope drainage requires inside ditches
Inside ditches	Ditch along inside of road	Carry runoff along road	May undermine cut slope and promote mass soil movement, requires cross-drainage
Cross-drainage culvert	Underground pipe or log construction	Drain inside ditches or natural waterway	Locate at low point of road, avoid open-top culvert
Water bar (turnout)	Low earth hump or log, 30° angle downslope	Disperse drainage from road	More frequent with steeper road grade
Broadbased dip	Wide drainage dip with reverse slope on road	Disperse drainage from road	Discharge onto undisturbed forest litter or stone, etc.
Stream crossings (bridges and culverts)	Minimize number of crossings, avoid channel changes, use riprap on approaches	Reduce effect of construction and erosion from approaches	Crossings are great threat to fish habitat
Vegetation planting	Seed grass, plant trees, brush on road edges and cut and fill slopes	Reduce erosion from exposed surfaces	Major element in protection of abandoned roads
Daylighting	Cut canopy of trees and brush to permit sunlight	Promote drying of roadbed and fills	Most useful for narrow roads

(Continued)

TABLE 4.—Continued.

Design feature	Method	Purpose	Remarks
Abandoned roads	Close access, remove bridges and culverts, install broadbased dips and water bars, vegetation	Avoid subsequent use and maintenance	Ideally, slope and reconstruct roads to avoid structural maintenance

[a]Based on: Lantz 1971; Larse 1971; Yee and Roelofs 1980; Swift 1984a, 1985, 1988; NCDEHNR 1989; Furniss et al. 1991; Maine Forest Service 1991.

Lantz (1971) prepared a handbook of guidelines for stream protection in logging operations, with emphasis on road design, construction, and maintenance. Larse (1971) presented a detailed outline for prevention of sediment from forest roads, including the major categories of planning, design, construction, maintenance, and post-construction operations. In planning, Larse emphasized the need to involve professional specialists from several disciplines—soil scientists, geologists, biologists, and others. He recommended that road locations should include natural yarding areas (benches, ridge tops) well away from streams, and that stream crossings be selected to minimize channel disturbance. He also recommended the option of "not-to-road" in some topographic conditions.

Larse (1971) gave many recommendations in road design: performing field-checks before beginning construction, minimizing excavations, constructing narrow roads, installing rock-surfacing or other surface cover including pavement, frequent culverts, and water bars. Recommended construction details included minimizing soil disturbance; allowing for good drainage during construction; suspending operations in heavy rainstorms, in high stream flows, or when excessive moisture conditions exist; and clearing organic debris from roads and embankments. The author emphasized the need to plan construction to avoid later mass soil movement due to unstable slopes. Maintenance included blading and shaping to preserve road profiles, clearing ditches and culverts, and removing slide materials. Post-construction operations included attempting to avoid conditions that might promote mass soil movements or landslides, reducing fill slope steepness with rock at slope bottoms, planting vegetation on slopes, and using rock, wire, asphalt, or mulch to stabilize road surfaces.

Larse's (1971) paper remains one of the most comprehensive guides published and is cited frequently. A later road planning guide (Yee and Roelofs 1980) depended heavily on Larse, adding recommendations for road stabilization with oil and chemicals (but also adding cautions regarding water quality), and culvert design for fish migrations.

Several long-term research programs have made great contributions to research on many problems regarding the effects of forest management practices on stream fisheries: Coweeta Hydrologic Laboratory, North Carolina (Swift 1985; Swank and Crossley 1988) (Figure 2, page 27); Alsea Watershed Study, Oregon (Moring 1975b; Hall et al. 1987) (Figure 3, page 29); South Fork Salmon River (Figure 5, page 32) and other central Idaho streams (Platts and Megahan 1975;

Control of erosion from logging roads is one of the most important preventive measures in timber-harvest practice. The exposed soil of cut and fill slopes is highly vulnerable to movement; in this photograph, the slopes of both the cut (on right) and fill (on left) have been backsloped and planted with grass, and a brush barrier has been installed at the toe of the fill slope. Coweeta Hydrologic Laboratory, North Carolina. (Photograph courtesy of USDA Forest Service, Southern Research Station.)

Megahan et al. 1991, 1992); and Carnation Creek Watershed Project, British Columbia (Hartman and Scrivener 1990) (Figure 6, page 34).

Most research on the effects of forest management practices in the southern states has been done at the Coweeta Hydrologic Laboratory, North Carolina (Swank and Crossley 1988). Among the earliest reports of biological damage from sediment originating in logging roads were those by Tebo (1955, 1957), who reported effects of sediment on trout streams at the then Coweeta Experimental Forest. Roads included excessively steep grades, no surfacing material, no culverts or water bars, and improper skid trails, which were downhill and often directly in tributary streams. Tebo reported severe reductions in stream benthos. He strongly recommended improved methods of skidding and road construction.

Many advances in the design and construction of logging roads were developed at Coweeta (Swift 1985, 1988). Swift (1988) presented a history of road-related research at Coweeta, beginning in the 1930s. Grass and mulch were used extensively in these early years—particularly on cut and fill slopes—because it was recognized that these slopes were major sources of sediment. In the 1940s,

Water bars on forest roads and recreational trails are frequently installed on steeper grades to direct water off the trail and onto the forest litter, where it can infiltrate into the soil rather than erode the trail or road.

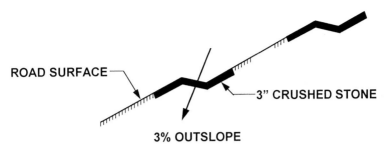

ROAD SURFACE

3" CRUSHED STONE

3% OUTSLOPE

FIGURE 18.—The "broad-based dip" removes water from the logging road and allows flow onto natural forest floor (from Kochenderfer 1970).

watershed treatment experiments were initiated using contemporary methods: skidding downslope or directly in streams; placing roads adjacent to or in streambeds; constructing roads without water diversions or protection by grass. Sediment input to streams was high. In the mid-1950s, however, road improvements were tested in experimental watersheds. Skid trails were avoided; skidding, mostly uphill, was by cable; streams and flow in intermittent channels crossed by roads were carried through culverts; roads were kept narrow; grass was planted on fill slopes and abandoned roads; and water bars, barriers, or logs were placed on roads. The result of using these new logging-road construction methods has been improved water quality.

A later improvement on the use of water bars was the "broad-based dip," which proved highly successful in directing drainage from roads (Kochenderfer 1970; USEPA 1993, Figure 18). The dip in the road includes a short reverse grade that prevents water from flowing down the road; water is diverted off the road in a wide, unconstricted flow and falls into undisturbed forest floor litter. Swift (1985) recommended, however, that in order for sediment leaving the road to remain on the forest floor, it should not cover a downslope distance greater than half the distance to a stream. The broad-based dip also replaced the inside ditch, which had the potential to undermine cut slopes and cause mass slippage. Swift (1988) also discussed other techniques, such as the use of grass cover and estimation of bridge and culvert sizes to handle floods. Grass surfacing on abandoned roads, on cut and fill slopes, and even on active road edges, was found to greatly reduce soil losses (Swift 1984a, 1984b). A manual of best management practice published by the North Carolina Department of Environment, Health, and Natural Resources (NCDEHNR 1989) addresses forestry practices specifically for the southern Appalachian Mountain region.

The Alsea Watershed Study compared three small experimental watersheds in western Oregon subjected to different logging practices: clear-cut, patch-cut, and an uncut control (Moring 1975b). Hall and Lantz (1969) had earlier discussed the effects of these different logging methods on sediment production and the consequent effects on reproductive success of coho salmon and rainbow trout. A later paper by Brown and Krygier (1971) addressed logging roads in the clear-cut and patch-cut watersheds and determined that the greatest increases in sediment production were attributable to roads in the logged watersheds, including an associated road slide and post-logging slash burning which exposed mineral soil.

The broad-based dip is used on logging roads to reduce erosion by removing water from the roadbed. The construction includes a shallow, broad depression followed by a reverse grade in the road, directing water off the road and onto the natural forest floor. (Photograph courtesy of USDA Forest Service, Southern Research Station.)

The long-term research in streams of central Idaho's highly erodible granitic batholith has emphasized the importance of proper road construction to avoid erosion. Early logging practices on the watershed of the South Fork Salmon River (SFSR), beginning in the 1940s, contributed immense quantities of sediment to the river, attributed largely to logging roads (Platts and Megahan 1975; Platts et al. 1989). However, Megahan et al. (1992), using a predictive sediment-yield model, estimated that by using present-day logging practices, the sediment inputs to the SFSR could have been reduced by up to 95%. Helicopter logging would have had a major effect, by reducing the distance of logging roads required from 46.7 km to 4.8 km, a reduction of 90%. This report documents the great negative effect that logging roads have as a producer of anthropogenic sediment and the concomitant loss of valuable stream fisheries.

The sediment-producing property of the central Idaho Batholith soils was pointed out early by Megahan and Kidd (1972) and others. They emphasized the need to address possible mass soil movements associated with roads. They also pointed out the need to treat surfaces of abandoned roads, because the soft granite weathered easily without treatment and produced sediment for many years.

Other streams in the central Idaho Batholith were also susceptible to sedimentation, especially from sediment originating in road-cut slopes, where exposed granite weathered quickly. Long-term sediment inputs occurred due to poor road design and placement, as well as to mass soil movements related to road construction (Platts et al. 1989). Megahan (1974) described the technique of planting deep-rooted trees to control slumping of road fills. Whereas grass would not prevent such mass erosion, the deep-rooted ponderosa pine, suitably treated with fertilizer and mulch, reduced erosion by 95% in 3 years. Megahan et al. (1991) presented methods of stabilizing road cuts and fills with vegetation seeding and planting, as well as techniques to evaluate treatments.

Fredriksen (1965, 1970) examined control methods in logging practices in the H. J. Andrews Experimental Forest (Figure 4, page 31). In this case, the clear-cut watershed was logged by skyline yarding (overhead cable) with no roads; the patch-cut watershed used high-lead yarding upslope but with roads; and the control was undisturbed. The clear-cut watershed (no roads) yielded suspended sediment concentrations 4.2 times higher than the control; the patch-cut watershed (with roads) yielded suspended sediment 39 times higher than the control. Bed-load sediment in the patch-cut streams was 178 times higher than in the control streams. The author emphasized the erodible nature of soils in the region and attributed the greatest sediment yield to landslides associated with roads. Maintenance of roads included surfacing with crushed rock, removal of slumps along cuts, clearing ditches, and grass seeding of cut and fill slopes. Two years after logging, suspended sediment yields from the patch-cut watershed returned to low levels only slightly higher than those of the control (Fredriksen 1965).

In the Carnation Creek Watershed Project, sediment eroded from logging-road surfaces was apparently not a serious problem; road surfaces were composed of hard, blasted rock, and even heavy traffic did not generate much sediment (Ottens and Rudd 1977). Sedimentation did occur, however, from eroded landslides. Because roads were constructed well away from the streams, few sediment fines reached experiments on salmonid spawning in Carnation Creek (Scrivener and Brownlee 1989). Their results contrasted sharply with those in the Idaho Batholith that involved soft granite (Megahan and Kidd 1972).

A few attempts have been made to analyze costs of special road construction and treatments to prevent road erosion, but these have met with difficulty. Ottens and Rudd (1977) examined a study road in the Carnation Creek project but obtained inconclusive results relating sediment from road construction to costs of damage prevention. Ward and Seiger (1983) developed a computer-based model to estimate sediment yields from various control measures and cost of control, using data from four New Mexico national forests. The program was deemed successful in providing cost estimates under specified conditions.

In Vermont, an extensive assessment of 78 timber harvest operations was conducted to determine compliance with the state's acceptable management practice guidelines and consequent soil erosion results (Brynn and Clausen 1991). Emphasis was placed on logging-road drainage techniques, skid trails and landings, and stream crossings. The authors concluded that although compliance was only partial, the overall result was minimal soil erosion. A major symposium, directed mainly at forestry and water quality in mid-south streams, emphasized mountainous regions of Arkansas (Blackmon 1985); several papers from the symposium are discussed in this volume.

Concern about the forest road as a sediment source has been expressed in other regions. Kochenderfer (1970) included many recommendations for design and maintenance of logging roads, based largely on studies at the Fernow Experimental Forest, West Virginia: locating roads away from streams; skidding logs uphill by cable; designing culverts, bridges, and water bars; adding broad-based dips with outsloped roadways; "daylighting" (opening the canopy over roads to promote drying of roads and fills); and installing filter strips along streams. In studies on a forest road in Oklahoma's Ouachita Mountains, Vowell (1985) indicated that sediment yields were proportional to the contributing catchment area upslope—useful information in determining the required size and frequency of roadway drainage devices.

Seehorn (1987), in a brief review of silvicultural practices and fisheries management in southern forests, recommended skyline cable or helicopter yarding on steep slopes. He also recommended buffer strips of about 50 m for steep slopes (70–80%) without brush barriers; with the addition of brush barriers, the width of buffer strips could be reduced to about 20 m. Modern timber harvest practices were reported for an Arizona mixed-conifer watershed: constructing vehicle stream crossings only when the streambed was armored; cutting in patches or selected groups; installing buffer strips at least 35 m wide between road and stream; and allowing tractor operations only along contours. These practices resulted in minimal overland flow and sediment production (Heede and King 1990).

In the southeastern Coastal Plains, loblolly pine plantations have become a major forest resource for the region. Plantations were developed on previously eroded crop lands and are now harvested periodically. Ursic (1991) assessed overland flow and sediment from harvesting operations, where planting and cutting were done on the contour, and where cutting was by strips (cut strips 20 m wide, uncut strips 10 m wide). Sediment increased after cutting (mainly in eroding channels) but decreased 4 years after harvest to levels that existed before cutting.

A guide for control of roadside erosion, primarily through revegetation, was developed for the specific conditions in British Columbia, where nutrient and soil pH problems are severe (Carr 1980). Structural measures and cutting practices to control erosion from timber-harvest operations in Maine were included in a recent handbook published by the Maine Forest Service (1991).

In studies of sediment transport from roads in the Ouachita Mountain forests, Arkansas, Miller et al. (1985) estimated sediment erosion from roads and delivery to streams. Only about 1% of sediment produced from roads reached the streams. This low rate was largely because the sediment-laden runoff was deposited on the natural forest floor rather than delivered directly to stream channels or natural drainageways. When runoff was discharged directly to channels and streams, however, about 50% of deposited sediment and 100% of suspended sediment reached the stream.

Whereas most published literature on the production of sediment from forestry management practices concerns the logging road, other factors have been reported along with preventive measures. For example, the cutting practice, particularly clear-cutting, is often suspect. However, the greater sediment generation from roads often masks the effect of cutting practices. Some early experiments on cutting methods indicted clear-cutting as a greater source than intensive selection (e.g., Eschner and Larmoyeux 1963), although no sediment-control measures were installed on skid roads in the clear-cut areas. More recent studies on clear-cutting also included the assessment of roads.

The erosion of streambanks, as a result of cutting or skidding directly in the riparian zone, may be second in importance to roads in producing excessive sediment. The breakdown of banks is difficult to avoid; the only way to eliminate this source of sediment appears to be by avoiding working in the riparian zone altogether (Meehan 1991b). Gregory and Ashkenas (1990) provided an extensive and detailed guide to riparian zone management in the Willamette National Forest, Oregon. The manual includes concerns for water quality, fish and wildlife, vegetation, timber, and recreation; it recommends practices intended to preserve many riparian values, including the control of sediment to streams.

The construction of major highways through forested areas, although not in the category of logging roads, is a potential source of sediment during construction (Brown and Mickey 1975). At risk in a North Carolina highway project were two high-quality trout streams with wild populations of rainbow trout and brown trout. Major erosion control measures were implemented: minimizing exposed area; providing plastic drains into brush barriers and settling basins; and promoting rapid revegetation using fertilizer and mulch.

In some cases, site preparation for replanting, particularly with mechanical means, can open and expose soils to erosion, as reported by Douglass and Goodwin (1980) in their experiments in North Carolina's Piedmont Province. However, rapid planting with grass reduced soil losses.

The method of skidding logs to access roads or yarding areas has long been recognized as a serious source of sediment, particularly when done by bulldozer. Kidd (1963) reported experimental treatments of skid trails (such as constructing slash dams and cross ditches, installing water bars, scattering slash on trail surfaces, and later seeding) with varying erosion responses. However, use of more recent techniques of uphill cable skidding (dispersing skid tracks rather than concentrating them with downhill skidding); skyline yarding from overhead cable (Oregon State University 1988); and helicopter logging (Megahan et al. 1992), would appear to almost eliminate the skid trail—and to a large extent the logging road as well—as sources of sediment.

Mining

Mining has been strongly indicted as a major source of pollution to streams of North America. The most acute mining effects have been from chemical and toxic pollutants stemming from release of acid waste (from coal mining that produces sulphuric acid); toxic metals such as chromium, copper, zinc, and others; and the use of highly toxic chemicals such as cyanide in gold ore processing. However, inorganic sediment production is an extremely important byproduct from several mining practices.

Placer gold mining, which directly intruded into streambeds or used hydraulic jets to sort gravel deposits near streams, was perhaps the practice that most seriously affected streams with excessive sediment. The major mining sources of sediment remain today as spoil piles resulting from surface mining, and tailings deposits resulting from ore processing, both of which are potentially subject to water erosion and transport to streams.

A particular form of mining waste is the sediment, especially clay, that is generated by placer gold mining in Alaska. Washwater from the sluicing process is run through settling ponds, but suspended sediment often passes through to contaminate the receiving stream. Treatment by passing through sand filters also removes some of the suspended sediment but not all. Chemical flocculants (e.g., lime, gypsum, or more recently, a water-soluble resin, polyethylene oxide) can be used to clarify turbidity, or remove suspended sediment, from the washwater. Settling basins may effect precipitation of the flocculates before releasing clarified water into streams (Weber 1986; Pain 1987).

The major means of preventing erosion and sediment production from mining wastes is containment on-site; revegetation is often used as a method of reclamation.

The literature on reclamation of mining wastes is large, including a number of books, many publications resulting from conferences and symposia, at least two primary serials, and many individual papers in related journals. Several natural resource agencies have been involved in problems and research on control of mining wastes, primarily in western arid and forested regions: U.S. Forest Service, U.S. Bureau of Reclamation, U.S. Environmental Protection Agency, U.S. Fish and Wildlife Service, British Columbia Ministry of Forests and Lands, and others. In 1973, the U.S. Forest Service initiated its Surface Environment and Mining program (SEAM), which included research, development, and application of solutions to problems of land impacts in the West. Many programs were developed, and resulting publications constitute a substantial contribution to the literature on this subject.

A sampling of books include those by Vories (1976), proceedings of a workshop on reclamation of western lands; Thames (1977), a comprehensive reference volume with contributions by many authors from the southwestern United States; and Schaller and Sutton (1978), proceedings of a 1976 symposium covering problems in diverse localities in the United States and Canada, sponsored by many agricultural and resource agencies and organizations. The U.S. Forest Service publications resulting from its SEAM program include Harwood (1979) and USFS (1979a, 1979b, 1980). The handbook of mining engineering (two volumes) published by the Society of Mining Engineers contains many valuable environmental contributions relating to prevention of pollution, including erosion control (Cummins and Given 1973a, 1973b). Many papers relevant to sediment control have been published in the two primary journals, *Reclamation* and *Revegetation Research and Reclamation Review*. Nelson et al. (1991) produced an important summary of mining influences on stream fisheries and mining waste control measures.

Spoil piles.—Spoil piles are the most important sources of sediment and present the greatest challenge to sediment control of all mining wastes.

A large body of literature is devoted to the reclamation of lands that surface mining has disturbed. Many techniques have been developed for the control of erosion from spoil piles (Table 5). Most research and development effort has been directed toward renovating the landscape to improve aesthetic and visual values, and to recreate the productivity of the land in a functioning ecosystem (Vories 1976; Nawrot et al. 1982). Streams or other water bodies often are not included in the landscape being reclaimed but almost all reclamation programs for spoil piles include erosion control techniques—measures that can serve just as well to prevent sediment delivery to local streams. Thus, many of the methods that have been developed chiefly to reclaim terrestrial values of a severely damaged landscape can serve also to eliminate or reduce sediment production and transport to streams.

Objectives of reclamation are mainly to serve some future use such as livestock grazing, agriculture, urban development, or public recreation. Most efforts are aimed at terrestrial rehabilitation or in many cases, the development of lakes and ponds for recreational fishing or commercial aquaculture. However, problems of erosion and sediment are often serious, especially in steep terrain or where excessively steep spoil piles have accumulated.

Reclamation for erosion reduction takes three major forms: (1) recontouring landscape features to original conditions; (2) installing water-retention structures; and (3) promoting revegetation. Recommendations for implementing these three erosion controls are taken largely from the publications edited by Vories (1976) and Schaller and Sutton (1978); the U.S. Forest Service's SEAM guides to vegetation (USFS 1979a), soils (USFS 1979b), and hydrology (USFS 1980); and the summary of mining effects and ameliorative measures by Bell and Payne (1993).

Recontouring the landscape. Although detailed techniques should be developed on a site-specific basis, several general methods appear to be common in reclamation of spoils:

- reservation of topsoil for final application, to provide a good growth medium for plants and better water infiltration;
- reduction of grades and lengths of slopes to reduce rapid runoff (less than 30% grade);
- avoidance of compaction of soil by heavy machinery as much as possible, to reduce rapid runoff (roughening of surface may be desirable, such as pitting and gouging);
- restoration by shaping and blading on the contour;
- installation of terraces on steep slopes;
- diversion of sediment-carrying water to level, nonsensitive areas (level spreaders);
- reduction of exposure times for erodible surfaces.

Techniques are more specifically described by Glover et. al. (1978) for humid regions (eastern areas), and by Verma and Thames (1978) and USFS (1980) for arid, western regions.

Water retention measures. Structural techniques should be designed and selected on a site-specific basis. Choices may be made from:

TABLE 5.—Sediment control on mining-disturbed lands—spoil piles and mine dumps.[a]

Method	Purpose	Remarks
Landscape recontouring		
Grading on contour	Reduce runoff on slope, promote infiltration	Leaves machinery tracks desirably on contours
Slope reduction, preferably <30%	Reduce rapid runoff	Steep slopes often the major problem
Reduce slope length	Reduce concentration of runoff	Needs initial planning; slope may be broken by benches
Avoid surface soil compaction, roughening with pitting and gouging	Slow runoff, promote infiltration	Compaction common with heavy machinery operation
Subsoil layering, compact layer below root zone	Retain water available to plants	Useful with coarse surface soils, maintain water-holding capacity
Reduced exposure time	Less erosion on exposed surfaces	Planning for less time between operations, rapid revegetation
Topsoiling	Enhance revegetation and moisture retention	Topsoil must be reserved in initial construction
Water retention		
Terraces	Slow runoff, promote infiltration of water	Most useful on steep slopes, often with reverse grade, may need special outlets
Benches	Break up long slopes, reduce runoff velocity	Terraces and benches may result in excessive water retention, causing landslides
Diversions across slopes, ditches, berms, water bars on roads	Remove water from slopes without excessive runoff	Water may be diverted to level, nonsensitive areas (level spreaders)
Grassed waterways, paved chutes, and pipe drains	Transport water from diversions, avoid gullying	Grass with low gradients <10%, paving or pipe with steep gradients or outlets from terraces
Sediment traps	Prevent fine sediment transport to waterway or stream	Usually small, temporary, often at toe of slope
Level spreader	Divert water and sediment to level area for infiltration and retention	Outlet to established vegetated area
Sediment basins	Permanent retention of water and sediment	Large, permanent pools— "last line of defense" to prevent off-site damage
Revegetation		
Site preparation	Gouging, pitting of surface on contour, use of reserved topsoil, amendments, organic residue	Provides for good germination and growth

(Continued)

TABLE 5.—Continued.

Method	Purpose	Remarks
Tillage	Loosen compacted soil, ripping, disking, harrowing	Provides aeration, water infiltration
Plant types	Forage grasses, legumes, forbs, shrubs, trees	Choice depends on geographical region, slope, future use
Seeding	Drilling (seeds in holes or furrows and covered), or broadcast (scatter seeds on surface); hydroseeding (broadcast mixture seed and water)	Native species often preferred, varies geographically; commercial suppliers or Plant Material Centers (USSCS)
Planting	Seedlings, transplants, root stock, cuttings, containerized stock	Plant when soil moisture optimum; most useful in harsh environments or steep slopes, for rapid cover
Mulches and amendments	Addition of organic materials— wood fiber, bark, manure, straw, sawdust	Promote aeration, nutrient retention; reduce wind erosion and moisture losses
Fertilizers and lime	Incorporated during tillage	Compensates for common infertility and low pH of spoils, particularly necessary for mine dumps containing ore residues
Irrigation	Artificial floodways, sprinkler systems, piping with trickle or drop application	Most important in arid regions, usually first year only for initial establishment

[a]Based on: USEPA 1976; Vories 1976; Thames 1977; Schaller and Sutton 1978—chapters by Bennet et al., Glover et al., Packer and Aldon, Thornburg and Fuchs, Verma and Thames; Harwood 1979; USFS 1979a, 1979b, 1980; Nelson et al. 1991.

- terraces and benches, often with reverse grades;
- diversions of water across slopes;
- grassed waterways;
- sediment traps at toes of slopes, and other grade-control and water-retaining structures.

Others are described and illustrated by Glover et al. (1978).

Revegetation. Establishing vegetation cover on spoil piles and other disturbed lands is the foremost method in all reclamation projects to restore the original function of the land, for agricultural development, and to reduce erosion and sediment production. Revegetation has thus been the most common and strongly emphasized technique for sediment control. Major publications include Vories (1976), Thames (1977), specific chapters in Schaller and Sutton (1978), USFS (1979a), and ASSMR (1985).

The selection of grasses, forbs, shrubs, and trees should be site-specific. Thornburg and Fuchs (1978) discussed regional criteria for plant selection in western dry regions that differ among the western Great Lakes, northern and central prairies, northern Great Plains, southern Great Plains, southern plains, southern plateaus, intermountain desertic basins, desert southwest, and California valleys. Bennet et al. (1978) discussed criteria for plant selection in eastern, humid regions, including the northeastern United States, middle and southern Appalachia, southeastern Canada, the northern Gulf states, and the eastern Great Plains. The major concerns in revegetation are plant materials, planting techniques, and plant species.

- Plant materials.—These include seeds, root stock, transplants, cuttings and sprigs, and container-grown stock (USFS 1979a). General classes of particularly useful plants for revegetation include forage grasses, forage legumes, and trees and shrubs (Bennet et al. 1978).
- Planting techniques.—As summarized by Vories (1976), Packer and Aldon (1978), and USFS (1979a), these include:
  site preparation;
  different types of tillage, partly aimed also at water retention (USFS 1979a);
  type of seeding, such as broadcast, drilling, hydroseeding (spraying seed and water slurry);
  timing of seeding (relating to season);
  use of mulches and fertilizers;
  special uses of seedlings and transplants;
  irrigation.
- Plant species.—The selection of plant species has been given great attention and species lists have been published for various regions and conditions (Thames 1977; Schaller and Sutton 1978, several chapters; and USFS 1979a). Factors influencing choices include many site-specific conditions, preference for native species and stocks, seeds in local topsoil, and cultured seed sources.

Tailings dumps and ponds.—Tailings generated either at mine sites or mill processing operations, often require restorative techniques similar to spoil piles when erosion of large dumps occurs (Table 6). In these cases, grading, structural controls, and revegetation techniques are similar to those outlined above for spoils. In the case of tailings ponds, which are developed where the processing of ores requires copious use of water for washing and grading, a slurry of tailings and fines is deposited into ponds (that sometimes cover huge areas) for filtering, settling, and permanent storage.

Another major difference between spoil piles and tailings ponds should be stressed. Spoils are the materials removed as overburden—soil, glacial till, sand, and rock—that do not contain ore but which overlie ore deposits and are removed to expose the ore. Spoils, therefore, do not usually contain appreciable amounts of mineral ore particles. Tailings, however, are the wastes generated by mechanical or chemical treatment of the ore itself and often do contain mineral particles and solutions, some of which may be highly toxic: asbestos-like fibers, heavy metals, uranium, iron pyrite (which produces sulphuric acid), or chemicals used in ore processing (Nielson and Peterson 1978). Tailings may also be either highly acidic

TABLE 6.—Sediment control on mining-disturbed lands—tailings ponds.[a]

Method	Purpose	Remarks
Tailings ponds as settling basin	Retain water, sediment, and tailings during ore processing	Need tight control of dam, spillway, outlets, and seepage
Control similar to spoil pile treatment, post-operation	Control sediment from accumulations in pond	Copious water used in processing, tailings contain ore residues
Diversion of excess water	Dewatering of pond accumulation to prevent erosive surface flow	May require level spreaders and basins
Dewatering of abandoned tailings pond	Post-operation reclamation	Dewatering necessary to cover toxic materials with inert soil below vegetation zone
Topsoiling	Provide suitable substrate for vegetation	Unlike reserved topsoil from spoil piles, topsoil must be obtained externally, preferably near original site
Revegetation of inactive ponds	Control surface erosion from settled material in abandoned ponds	Toxic substances may retard plant growth; lime and fertilizer often necessary; unvegetated surface may also cause serious dust problems
Permanent settling basins below abandoned ponds	Removal of fines	Require periodic dredging; may be used for fish and wildlife depending on water quality

[a]Based on: Schaller and Sutton 1978, chapter by Nielson and Peterson; Harwood 1979.

or alkaline, thus requiring treatment to produce pH levels that will support vegetation (Thames 1977; Yamamoto 1982).

The toxicity of tailings may be a problem for chemical pollution control, but it may indirectly affect the generation of inert sediment as well. Toxic conditions in a tailings pond can result in failure to establish vegetation that can be intolerant to the specific toxic substance in the root zone of the plants. Thus, revegetation attempts may fail to provide the desired erosion control (Harwood 1979).

Major techniques of sediment control on abandoned tailings ponds include:

• recontouring to promote water retention and diversion of excess water to prevent erosion;
• promoting revegetation to hold surface soils and prevent erosion.

Another difference between the tailings pond treatment and spoil piles treatment is that the tailings pond often must be dewatered first so that toxic materials and water can be covered by inert material well below the root zones of the plants used.

TABLE 7.—Sediment control on mining-disturbed lands—sand and gravel excavations.[a]

Type of excavation	Control method	Remarks
Instream dredging	Rock gabion to halt head-cutting	Gabion is extreme measure, may form waterfall and obstacle to fish migration
	Complete prohibition	Instream excavation causes extreme damage, even with mitigative practices and structures
Floodplain excavations	Settling ponds, with outlets	Removes coarser material only, fines returned to stream
	Outlets with structures to avoid streambank erosion, provide filtering, buffer strips	Export of fines from gravel excavation and washing major impact on streams
	Avoid new channels	Erosion of new channels may cause extreme siltation of receiving stream
	Settling ponds, with no outlets	Infiltration of water provides filtering of fines
	Complete filtering and recycling of water used in processing	Minimum or no impact to stream
Abandoned operation	Backfilling of excavation and revegetation	Return landscape to original condition
	Permanent ponds (no outlet), managed for fishing or other recreational use	Requires continued public management

[a]Based on: Blauch 1978; Woodward-Clyde Consultants 1980b; Kanehl and Lyons 1992.

Reclamation of active tailings ponds is not possible until mining ceases, but some measures may be used to lessen later generation of sediment:

- maintenance of the integrity and capacity of the dam;
- design of pond to avoid possibility of overtopping;
- prevention of erosion of dam spillway or underground outlets;
- installation of a series of small sediment settling basins downstream from the pond in the event of seepage (Cummins and Given 1973b).

Mining sand and gravel.—Extraction of sand and gravel and other earth materials is widespread; they may be the most common of mined materials (Starnes 1983, 1985). Yet the environmental influence of this form of mining appears to have received very little attention. Less information appears in the literature on sediment pollution control methods than for other forms of mining (Table 7).

In a comprehensive symposium on reclamation of mined lands (Schaller and Sutton 1978), a short chapter on sand, gravel, and rock mining by Blauch (1978) included little information on sediment control measures relative to other chapters dealing with mineral ore mining. Blauch suggested that sand and gravel pits excavated below the water table can be drained and dewatered, and then backfilled and revegetated, or they may be impounded to create recreational water bodies. He cautioned that streams flowing nearby or through the excavation site would require settling basins to maintain water quality.

The studies by Woodward-Clyde Consultants (1980a) on gravel mining operations in Alaska produced a manual of guidelines to protect against streambed degradation, including recommendations to avoid instream work, depths that created permanent pools or new channels, and disturbance to streambanks. They also recommended ponds for settling and filtering (i.e., with no outlet) to avoid elevated suspended sediment concentrations downstream (Woodward-Clyde Consultants 1980b).

Kanehl and Lyons (1992), after extensive studies of sand and gravel mining impacts in Wisconsin streams, concluded that a great void exists in the literature regarding monitoring and research on effects of sand and gravel extractions from streams and adjacent floodplains. They recommended elimination of in-stream sand and gravel mining. In the absence of such prohibition, however, they made strong recommendations relating to the study of control measures such as bank stabilization, revegetation, buffer strips, influence of connected floodplain pits, devices to control headcutting, and recycling of water from gravel washing operations.

Urban Development

Most available information about control of sediment from urban sources has originated in the densely populated area of the eastern United States. Sediment production by urban development may be small in terms of continental area, but effects on nearby streams can be severe (Wolman and Schick 1967). Rates of soil losses from exposed areas in construction sites, especially areas left open for long periods, have been reported many times higher than those from either forest or agricultural sources (Wark and Keller 1963; Yorke and Herb 1978). Many guides and handbooks with detailed methods to control sediment have been published (Table 8). Preliminary planning for permanent structures assist in providing sediment-free maintenance of residential and industrial urban areas. Selection of structure location, integration with local topography and nearby streams, permanent grassed waterways, road bridges and culverts, ponds (perhaps useful for recreation), as well as plans for post-construction operation, all should be considered.

Preventing overland runoff and sediment from entering a construction site—by using water diversions around areas planned for construction—may forestall many problems of handling sediment on the site (WDNR 1992). Losses of sediment from on-site areas, however, can generally be prevented or reduced by some means of containment (Wolman 1964; USSCS 1973; WDNR 1992). Retention of water and sediment on the construction site involves two general

TABLE 8.—Methods for the control of erosion on urban construction sites.[a]

Method	Purpose	Remarks
Off-site		
Water diversions across slopes, channels and berms, runoff directed to non-erodible areas or detention basins	Prevent water and sediment from entering site during construction	Upslope from construction site
Temporary (12 months)		2-year storm frequency, vegetation seeded or sodded, straw filters
Permanent		25-year storm frequency, vegetation on side slopes, permanent vegetation in channels, or (if >3% grade) rock riprap, plastic sheet, culverts, or pipes
Inlet filter	Filter and retain sediment at point of entry to construction site	Usually temporary, or permanent if surface area is reserved for control use
Level benches and terraces	Slow runoff and promote infiltration	May be combined with diversions, especially on steeper slopes
On-site		
Initial land grading with gentle slopes, berms and terraces, retention walls, detention dams	Prevent water and sediment runoff from site during construction—containment of sediment produced on-site	Containment methods with many variations of filtering and settling structures, mostly temporary
Settling traps and basins	Interdiction of sediment	Temporary—dewater after construction
Sediment barriers—hay, straw, or fiber filter fences	Filter and retain sediment from exiting water	Often used at base of slopes
Design of access roads—water bars and humps, drainage culverts, etc.	Reduce roadbed erosion	Generally, same practices as for logging roads (see Table 4)
Short exposure—minimize time between soil-disturbing activities; temporary sodding or seeding; fiber mat covering	Reduce erosion from disturbed soil	Preconstruction schedule planning required
Vegetation Temporary—annual fast-growing plants or sod Permanent—perennial plants, bushes, trees	Eliminate erosion on graded slopes, waterways, fill slopes, bridge approaches	Essential element in all urban erosion control
Level spreader—grade broad, undeveloped area, vegetated	Allow runoff water and sediment to disperse over large, level area	Water to infiltrate, sediment to remain in place

(Continued)

TABLE 8.—Continued.

Method	Purpose	Remarks
Post-construction		
Permanent grassed drainageways	Conveyance of moderate runoff without erosion	On gentle grades, low velocity and discharge, 10-year storm frequency
Surfaced channels—rock or concrete	Transport of stormwater discharge	Intermittent, high-velocity discharge, combined with energy dissipators (rock filters, baffles)
Sediment basins—permanent ponds	Trap and retain sediment leaving developed site	May need periodic dewatering and dredging; ponds may be managed for fishing or other recreational uses
Streambank stabilization—bank grading, riprap (see text on subject)	Prevent sediment production by streambank erosion	Natural stream discharge may increase after urbanization
Steep slope stabilization—rock and gravel, check dams, energy dissipators	Prevent erosion on steep slopes that cannot be protected by vegetation	Such steep slopes normally should be excluded in planning process

[a]Based on: Hansen 1968; USEPA 1973b; USSCS 1973; Goldman et al. 1986; WDNR 1989.

concepts: holding water on-site with various surface controls and reducing the duration of exposure.

Initial land grading in various forms serves to retain runoff, either temporarily or permanently, and assists in the containment of water and sediment on the site. These measures include grading for gentle slopes, creation of berms or terraces on contours, harrowing to pulverize soils along contours, retention walls, and small detention dams to create temporary settling basins. Use of hay or straw bales and fiber filters, planned to temporarily surround land-disturbing activities or to temporarily cover exposed soil, also assist in containment. Temporarily diverting runoff to level spreaders (shallow, flat basins) permits sediment to settle out and water to permeate the surface. Other measures intended to retain runoff, rather than expedite flow, may be fitted to local topography.

A major structural measure is seeding for temporary vegetative cover during the time of construction, using mulch (hay, straw, or fiber matting) for protection during seed germination. Small grains (e.g., rye or oats) and legumes are commonly used. Seeding immediately after soil disturbance reduces exposure times. Where rapid cover is required, especially on steep slopes, sodding is preferable to seeding. Temporary measures that may substitute for vegetation in shorter durations include the use of mulching and fiber matting to prevent runoff.

A special problem exists where urban development occurs in regions with silt and clay, e.g., the red-clay area of the Piedmont Province in the eastern United States (Lemly 1982a). Small particles (<0.04 mm) are very difficult to control with the usual filtering techniques. Lemly conducted experiments on the effectiveness of the standard erosion control techniques of vegetation seeding, chemical soil

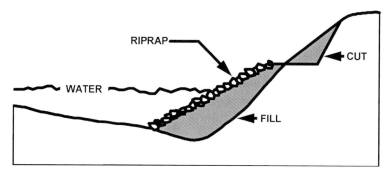

FIGURE 19.—Construction of upper and lower banks from an initially high, steep, eroding streambank (from Hansen 1968).

binders, straw, and jute netting, and concluded that multiple treatments (several of the listed treatments in series) were necessary to effectively filter the finer sediments.

Proper design of access roads (as in any other practice that generates erodible sediment, such as logging) can assist in retaining runoff, particularly from temporary roads used by heavy construction equipment. Back-sloping, use of humps or water bars, and temporary culverts assist in controlling runoff from construction roads.

Details of control practices, structures, and permanent maintenance, including model sediment control ordinances, are given in many guides and handbooks available from local and state agencies. Goldman et al. (1986) is the major handbook for general use in urban sediment control; regional examples are those for urban Maryland (Wolman 1964), New Hampshire (USSCS 1973), and Wisconsin (WDNR 1992).

Streambank Erosion

Measures to prevent or control streambank erosion were some of the earliest sediment control techniques to be developed. Slumping and eroding streambanks in agricultural areas were readily observed as having been caused by livestock trampling or stream-edge cultivation.

Principles of control recognize both an "upper bank" above the zone influenced by high water or floods, and a "lower bank" next to normal stream water levels and scoured by spates or floods (Hansen 1968) (Figure 19). The upper bank would not normally require structural protection, although runoff might be severe due to a steep slope. The lower bank is the zone most susceptible to erosion and may require special structural protection. Because of the importance of bank slumping, reduction of slopes on both upper and lower banks is usually necessary (Table 9).

Two general goals are pursued to prevent erosion of the lower bank: (1) increase the resistance to erosion of the lower bank with some kind of surficial treatment (riprap); and (2) reduce the energy of water potentially scouring the lower bank (Gregory and Stokoe 1981).

TABLE 9.—Methods for reducing streambank erosion.[a]

Treatment zone	Method	Remarks
Lower bank (directly impacted by stream flow)	Reduction of water energy, instream structures	May not be compatible with aesthetic, boating, or fishery goals
	Slope reduction	Requires less slope with flow maximum, gradient, storm frequency
	Riprap—rock, brush	Most common practice (trash not recommended)
	Revegetation	Grass turf, brush, trees with strong root system (e.g., willows)
Upper bank (above high water zone and floods)	Slope reduction	Usually less slope than lower bank
	Terrace at toe of slope	Reduces water runoff velocity, useful for foot traffic
	Revegetation	Turf, trees
Riparian zone	Fencing	Eliminates foot traffic and livestock grazing; may be damaged or eliminated by streamside structures
Watershed influences	Erosion control in uplands	Basic practice essential to good streambank stabilization practice

[a]Based on: White and Brynildson 1967; Hansen 1968; CDFO 1980; Gregory and Stokoe 1981; Claire and Storch 1983; Platts and Nelson 1985a.

Normally, the first of these would be most commonly employed. The second, reduction of water energy, entails instream deflectors, retards, or brush, log, and rock barriers. These structures, however, usually are avoided for aesthetic, boating, and fish management reasons.

Resistance of the lower bank can be increased by use of several techniques: (1) sloping back at an angle in accordance with the degree of bank material cohesion; (2) surficial covering with rock riprap (most common), logs and brush, or other non-erodible material (Formerly, used tires, wrecked auto bodies, mattresses, etc., were sometimes used for this purpose but for obvious aesthetic reasons, and because some of these may release toxic contaminants, their use is strongly discouraged and in many areas is illegal.); and (3) revegetation to produce a natural appearance and further improve bank stabilization.

Further recommendations include structural measures to stabilize streambanks. Plantings of trees or grass turf are often recommended for the upper bank above high water (CDFO 1980); use of a tough turf at streamside was highly recommended by White and Brynildson (1967) and Hunt (1988) for low-gradient

Development of a streambank erosion control project in South Branch Root River, a popular trout stream in southeastern Minnesota. Top: Severely eroding bank at popular access point, 1972. Bottom: Identical location after construction of bank stabilization, photo taken in 1994.

midwestern streams. A number of habitat alteration guides include principles and methods aimed at reducing streambank erosion.

Many authors who report stream degradation by livestock grazing in western rangelands have been concerned with the destabilization of streambanks (with consequent sediment production) from trampling by animals. Fencing against the use of the riparian zone is the common practice of control and is most effective when combined with some revegetation, such as willow plantings, to anchor the banks (Claire and Storch 1983). Platts and Nelson (1985a, 1985b) reported the protection of streambanks with livestock exclosures but cautioned that other factors, such as the length of stream reach protected, also need to be considered if improvement of the fishery is an objective.

Finally, while listing and describing some bank stabilization techniques, Lines et al. (1979) cautioned that these measures often displace the riparian zone itself, with its many values to both wildlife and stream fish. They argued furthermore that structural techniques at streamside do not correct the original sources of the problem, such as poor land use practices in the watershed.

Continued monitoring and management of installed erosion-control systems are required on agricultural, forested, and mined lands. Landscape use obviously dictates general management practices, but monitoring with continuing attention to erosion and sediment prevention must be included to sustain the quality of a resource and its cultural applications, whether they are agriculture, industry, or wildland recreation.

The elimination of excessive erosion from cultivated, logged, or mined lands, or from any other source of anthropogenic sediments that affects stream fisheries, will continue to be a major concern. Watershed problems undoubtedly require solution before, or concurrently with, streambank erosion corrections aimed at sediment control in an ecosystem or holistic approach.

INTERDICTION

Interdiction as a method of sediment control is taken here to mean intercepting and retaining sediment in transport between the location of its origin and a stream. Two principal means have emerged: (1) buffer strips of vegetation, usually in the riparian zone along streambanks, to filter and retain sediment as part of the permanent riparian soil structure; and (2) sediment traps (settling basins) constructed either temporarily on land during a surface-disturbing operation or instream as permanent installations. The latter require periodical sediment removal by dredging.

Buffer Strips

The buffer strip (buffer zone, filter strip, leave strip, other terms) was one of the earliest techniques employed to control sediment in forest logging operations (Trimble and Sartz 1957; Haupt 1959). The buffer strip in its simplest form is a zone of natural vegetation left uncut along the sides of a stream in a logged watershed (Hall and Lantz 1969; Lantz 1971; Newbold et al. 1980; Culp and Davies 1983; Erman and Mahoney 1983; Kochenderfer and Edwards 1991). The zone may be modified by additions such as brush barriers, logs, and rocks (Hartung and Kress 1977; Swift 1986).

The buffer strip may be used in many kinds of applications other than forestry where land-disturbing activity generates sediment, for example, with agricultural cultivation (Roseboom and Russell 1985; Whitworth and Martin 1990; Delong and Brusven 1991); livestock grazing (Elmore 1992); mining (Glover et al. 1978; Kanehl and Lyons 1992); and urban construction (USEPA 1973b; Hartung and Kress 1977). Lowrance et al. (1985) emphasized the importance of a healthy riparian zone in filtering several types of nonpoint-source pollutants from agricultural areas, including sediment.

A recent USEPA report (Whitworth and Martin 1990) documented the results of agricultural buffer strips left in place as part of the Conservation Reserve Program (CRP), a program that increasingly includes the use of filter strips in agriculture. Comparing streams in agricultural areas of Indiana and North Carolina, the authors reported that streams with filter strips had higher density and taxa richness for macroinvertebrates, fish species diversity, fish density, and index of biotic integrity (Karr et al. 1986). Whitworth and Martin (1990) urged expanded use and research evaluation of filter strips in federal CRP applications.

Karr and Schlosser (1977) developed an early review of the use of nearstream vegetation (i.e., the riparian zone) to reduce transport of sediment to streams. They concluded that overland flow was by far the major source of sediment to streams and that nearstream vegetation strips could effectively reduce such inputs by filtering. Vegetated roadside ditches also filtered coarser sediment eroded from graveled roads, particularly when ditches were of low gradient (Bilby 1985); the finest sediments (<0.004 mm), although not entirely filtered in the ditches, were washed away in a receiving stream during high discharges.

Considerable effort has been allocated to planning and designing buffer strips to ensure proper function. The choice of width was one of the major concerns in the earliest literature. Trimble and Sartz (1957) measured the distance of sediment flow from logging road culverts. They developed a table of recommended filter strips based on the slope of land between road and stream, ranging from 25 ft at 0% slope to 165 ft at 70% slope, for general forest situations. Recommendations in the table were doubled, however, where stream water was intended as a municipal water supply. Haupt (1959) emphasized that the "protective strip" should be wide enough to dissipate the sediment load. He provided directions for calculating required buffer widths based on cross-ditch spacing on logging roads, slope obstructions, direction of slope exposure, and other factors. Hartung and Kress (1977) provided a table of strip-width recommendations based on logging areas versus municipal areas and percent slope. Recommendations ranged from 25 ft in logging areas with 0% slope, to 450 ft in municipal areas with maximum slope to protect water quality.

In one of the most extensive studies of the effect on invertebrates and buffer strips in logging operations, 62 northern California streams with different buffer strip widths were sampled for invertebrates. Four categories of streams were sampled: no buffer strips, buffer strips less than 30 m wide, buffer strips 30–60 m wide, and controls (no logging) (Erman et al. 1977; Roby et al. 1977). The authors reported benthos diversity to be the same in controls and in streams with wide buffers, but lower diversity in streams with no buffers or narrow buffers. Streams without buffer strips still showed lower diversity after 10 years. Although the 30-m division was arbitrary, this choice appeared to separate distinct categories of effects by an approximate effective width.

Swift (1986) reviewed the problem of determining appropriate buffer strip width, based on comparisons among previously published data. Based on his research at the Coweeta Hydrologic Laboratory, North Carolina, however, Swift concluded that narrower widths could serve as well in certain conditions, such as grassed roadbeds and fill slopes, and with use of brush barriers and erosion control measures on roads. He presented recommendations for widths to only 64 ft at 80% slope when graveled roads, grassed banks and fill slopes, and brush barriers were included. However, Swift cautioned that without these features, the indicated widths should be increased.

Other design features were described by Steinblums et al. (1984), who pointed out the necessity for buffer strip stability. They emphasized that wind damage was a critical factor in strip stability and that if blowdowns or other damages from further logging or disease occur, the resultant debris can load stream channels and cause bank erosion during high flows. Factors that affect wind damage, in addition to strip width, included tree size and species, canopy density, slope in the direction of the prevailing wind, local topography including nearby ridges, slope of streambanks, and others. Delong and Brusven (1991) developed a classification system of riparian habitats that receive runoff from agricultural fields. Their analysis used parameters such as riparian slope and width, height of vegetation, and land use to provide the capability to predict the susceptibility of stream reaches to damage from nonpoint-source pollution.

The latest and most comprehensive review of use of riparian zones (i.e., buffer strips) to protect stream fisheries is that by Moring and Garman (1986). They analyzed the literature from the standpoint of the changes incurred when the riparian zone is removed by logging. Increased sedimentation was a major concern, and effects also included increased stream flow (including floods), increased streambank erosion, higher turbidity, elevated water temperatures, and decreased allochthonous leaf litter. Resulting biological effects included decreased aquatic invertebrate productivity and terrestrial invertebrate drift, reduced fish reproductive success from sediment deposits on eggs and developing embryos, and loss of fish-rearing habitat. Moring and Garman concluded that, without buffers, recovery from stream damage can be protracted and might never occur.

Moring and Garman (1986) also included results from their own study on the East Branch Piscataquis River, Maine, where much of the riparian zone was removed in a logging operation. Higher suspended sediment levels occurred, greater amounts of fines were deposited in pool gravels, temperatures were elevated, brook trout were eliminated, and density of nongame fishes increased. They recommended buffer strips of at least 75 ft, although slopes in the watershed were not reported.

Obviously, the width of buffer strip required to protect a receiving stream will vary according to local conditions but a range of about 50–300 ft would appear to constitute a general guideline. Published recommendations for buffer strip widths vary widely, calculated as they have been from a variety of timber-harvest conditions. Protection of the riparian zone and elimination of streambank erosion are the main objectives, and for those purposes Meehan (1991b) has made perhaps the best recommendation: to eliminate logging activity in the riparian zone altogether.

Hillside slope in the watershed is probably the most critical factor in determining the effect of buffer strips, but land use is also of great importance. Most applications have been in forestry operations, usually to intercept eroded sediment from logging roads, but any other activities that disturb land surfaces also may require this interdiction technique. The transport of sediment from urban developments, particularly with their large proportions of land area covered with impervious surfaces, may necessitate buffer strips of greatest width.

With all of the research efforts, experiments, trials, and evaluations of buffer strip width and composition, the major control measure involving interdiction of sediment between its source and the stream is the avoidance of any surface-disturbing activity of any kind in the riparian zone. The best management practice to prevent severe stream sedimentation is clearly the total elimination of agricultural cultivation, livestock grazing, timber harvest, mining of all kinds, and any residential and recreational development, from highly sensitive riparian zones and floodplains. Only this broad policy will provide the maximum protection from anthropogenic sediment to stream values of all kinds.

Sediment Traps

The most efficient device to interdict eroded sediment flowing in a channel, particularly bed load, is the sediment trap (sand trap, settling basin, detention dam, and other terms). Its usefulness derives from the fact that sediment moves only in a current; when current velocity drops, as in a pool or impoundment, sediment falls out to form bottom deposits.

The sediment trap can take several forms. Small on-land temporary impoundments may be employed to retain sediment during urban development, highway construction, or logging operations. Farm ponds in line with natural drainageways, or a series of basins below impoundments of ore tailings, can interdict overland flow. Large, permanent dams constructed for other purposes can also serve to collect sediment. Instream excavations or impoundments in small streams can effectively protect downstream fisheries.

Sediment traps can be categorized into several types. Probably the simplest and oldest in use are small traps constructed on-land, either as impoundments or excavations below mining spoil piles and tailings ponds (Tryon et al. 1976; Harwood 1979; Bucek 1981); at the toe of fill slopes below logging roads (Haupt 1959; Swift 1986); around urban construction sites (USEPA 1973b; Chen 1975; Baumann 1990; WDNR 1992); and in other operations that involve disturbance and exposure of land surfaces. These structures may be rudimentary and temporary, for short-term use only while the mining, logging, or construction operation proceeds, and need no further maintenance.

Other basins may be constructed with a slightly more elaborate design intended to function for longer periods, often in a series of basins (operating more efficiently than a single, larger impoundment of equal volume). Examples include those draining from ore tailings ponds (Cummins and Given 1973b) and sand-and-gravel dredging operations (Newport and Moyer 1974; Kanehl and Lyons 1992). These basins may require later dredging to renew their capacity or may be left as permanent features after they are filled and reclaimed with vegetative cover. The Wisconsin best management practice handbook (WDNR 1992) includes many construction details for temporary and semi-permanent sediment basin designs, some of which may be maintained for aesthetic purposes, fish ponds, skating rinks, or other recreational uses.

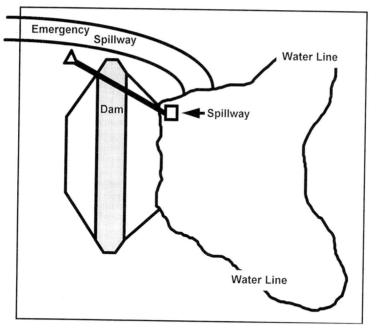

FIGURE 20.—Outline plan of "farm pond" used to interdict overland flow of runoff (from Allen and Lopinot 1968).

Farm ponds, although ordinarily constructed for livestock watering or recreational fisheries, may also serve as sediment traps (Beasley et al. 1984). The construction of a farm pond by damming an active stream is not recommended, especially if fish production is desired (Allen and Lopinot 1968), so the sediment trapping objective would be directed at overland rather than in-channel flows (Figure 20). Instead of damming a stream, therefore, an impoundment may be formed by damming at the lower end of a ravine or other natural depression in the landscape, where topography forms a constriction suitable for a dam. The pond may be perched above the water table, in which case rainfall will be the source of water. If necessary, the pond bottom may be treated with impermeable clay, such as bentonite, to prevent leakage (Gablehouse et al. 1982).

Alternatively, a farm pond may be formed by excavation, in which case groundwater probably will be the principal source of water, and the pond's water surface will be an expression of the water table (USSCS 1982). The excavated pond may be more expensive to build; it will also lack the capacity for spillway control and periodic drainage, which are advantages for management of the pond for fish production. Depending on the priority of purposes for the pond, accumulated sediment may have to be dredged periodically, whether constructed as an impoundment or as an excavation, or allowed to accumulate through the design life of the pond and then be reclaimed by revegetation.

When the purpose of the pond is principally for livestock or fisheries, sedimentation should be prevented by erosion-control measures in the watershed, buffer strips around the pond or other practices. The American Fisheries

Society, North Central Division, produced a guide to management of small impoundments for various fish species (Novinger and Dillard 1978).

Instream structures include both impoundments and excavations. The observation of higher productivity of invertebrates and fish below large reservoirs is common in the literature (MacFarlane and Waters 1982; Parker and Voshell 1983; Rader and Ward 1989). Observations of higher productivity have been usually attributed to increased organic seston and plankton originating in the reservoir; little consideration has been given to sediment factors. However, impoundments also remove large quantities of sediment by deposition, and the resulting clearer water downstream promotes increased productivity (Patrick 1976). Highly productive, cool, tailwater trout fisheries have been developed downstream from large dams with hypolimnetic discharges in previously warm, turbid rivers such as the Green River, Utah; Bighorn River, Montana; and San Juan River, New Mexico (author's observations).

Large reservoirs, however, are destined to fill with sediment and then to become useless for any function. Although it may be possible to decelerate the rate of sedimentation, filling is nevertheless inevitable (Simons 1979). Therefore, the function of large dams and reservoirs for purposes of sediment interdiction should be viewed with caution, because the possibility of renovating the reservoir by removing the sediment is probably nil. Other possible negative effects include greater sediment deposits upstream from the reservoir through aggradation and—with clearer water downstream from the dam—increased sediment transport capacity and consequent greater streambank erosion.

Seldom are large rivers impounded specifically for purposes of reducing sediment downstream. However, a multiple-purpose dam, installed on the North Branch Rock Creek, Maryland, included the purpose of water quality improvement (Herb 1980). Studies made by the U.S. Geological Survey showed an overall efficiency of sediment removal of 95% (136,000 tons entering in the period 1968–1976, and 5,910 tons leaving the impoundment), and an attenuation of a storm-generated sediment peak of more than 99% (4,880 tons/d upstream versus 33 tons/d below the dam).

The most notable example of a large dam and reservoir built specifically to interdict sediment and protect a stream fishery is the Buckhorn Dam on Grass Valley Creek, California (USDI 1990). Grass Valley Creek is a tributary of the Trinity River (Klamath River basin), which had reduced populations of steelhead trout and coho and chinook salmon, in large part due to heavy sedimentation from this tributary. Projected design life was 50 years; the reservoir was planned to trap about one-fourth of the annual sediment load of 170,000 yd^3 in Grass Valley Creek. In addition to the dam and reservoir, other sediment-reduction projects were planned for other tributaries. Other pools in Grass Valley Creek and the Trinity River were constructed to trap additional sediments; the pools were to be dredged periodically. The Buckhorn Dam project is solely a sediment-trapping facility, probably the only one of its size constructed with this specific objective.

On a smaller scale is the excavated sediment trap installed in small or medium-sized trout streams to protect downstream reaches. Nobel and Lundeen (1971) developed a cost-benefit plan to reduce sediment in the South Fork Salmon River, Idaho, heavily sedimented by past logging operations. Among a number of alternative methods, including road stabilization, skid road closures, and in-stream debris removal, they concluded that "channel debris basins" (i.e., sediment traps) would be the most economical and efficient in sediment removal.

Hansen (1973) suggested that the excavated basin was superior to an impoundment because the impoundment would tend to cause undesirable aggradation upstream. He further pointed out that an excavated basin can be used where gradient is relatively flat. Drawing upon experience with non-fishery-related sediment basins, Hansen presented means of designing physical dimensions of proposed sediment traps. Whereas such basins trap mostly coarser particles (e.g., sand), they often allow transported fines to pass through. Periodic dredging is required, and proper disposal of spoils is necessary. Hansen also cautioned that the resultant pools may pose a hazard to wading anglers.

Later, Alexander and Hansen (1983) and Hansen et al. (1983) described a sediment basin application in Poplar Creek, Michigan, that was specifically designed to trap and remove sand from the heavily sedimented stream. The authors pointed out that such basins trap moving bedload sand that, even at low levels, may be severely damaging to trout populations but may be undetectable by suspended sediment samplers. Their sediment basin in Poplar Creek removed 86% of the bed load, after which brown trout and rainbow trout numbers increased: by 40% for small and young trout, by 28% for total trout production; trout production in a control section remained constant.

In a similar sediment-trap installation in Valley Creek, Minnesota, annual production by hydropsychid caddisflies in a stony riffle downstream from the basin was sixfold higher than production in an upstream, sedimented section (Mackay and Waters 1986). No difference in seston quantity was observed between upstream and downstream sites. The authors postulated that the difference in streambed sedimentation accounted for more favorable hydropsychid attachment sites (stones and epilithic moss) at the downstream location. This basin was a combination excavation and impoundment and was dredged periodically. It successfully protected about 1 km of alternating pool-and-riffle habitat of high trout productivity.

The sediment trap, especially when installed in a small stream, is a highly efficient management tool to improve or protect the fishery. No doubt it represents intense management, but it is a good obvious interdiction technique for small but highly valued stream fisheries.

RESTORATION

In a broad context, "restoration" is defined as " . . . the re-establishment of predisturbance aquatic functions and related physical, chemical and biological characteristics" (NRC 1992). The principal concept here is "functions," that is, the recovery of biological communities in relation to their environment. The National Research Council (1992) lists such restorative procedures as change in land use, protection of the riparian zone, reintroduction of fish species, and others.

In a similar vein, Gore (1985a) described river restoration as the process of "recovery enhancement," that is, the design and placement of habitat features that attract colonizing biota by natural means. Herricks and Osborne (1985) list four

The sediment trap installed in a small trout stream is a highly effective device to interdict sediment and protect downstream reaches. Top: A small reservoir created partly by impoundment and partly by excavation decreases current velocity and causes settling of silt and sand. Bottom: A tongue of sand—visible in the upper portion of the reservoir— gradually moves down to eventually fill in the pool. Periodic dredging is required. This sediment trap resulted in much higher annual production of stream insects downstream from the dam. Valley Creek, Minnesota.

general principles of water quality restoration in polluted streams and rivers: isolation, removal, transfer, and dilution of the polluting material. Gore and Bryant (1988) further discussed these four measures, emphasizing the necessity of a holistic approach to the restoration of disturbed areas. A burgeoning literature on recovery theory currently deals with both natural and human-assisted return of viable stream communities after disturbance; this approach is exemplified by the comprehensive USEPA symposium edited by Yount and Niemi (1990).

A definition of "restoration" is used herein to apply to removal of sediment from a stream, or stream reach, usually by some means either of flushing downstream or dredging—that is, by either transfer or removal, in the sense of Herricks and Osborne (1985). Only physical restoration is thus implied—not necessarily a subsequent recovery of ecological function. With respect to sediment, examples of all four means might be: isolation, a permanent deposit behind a high dam; removal, dredging sediment for distant disposal; transfer, movement of sediment downstream; and dilution, transfer and redeposition downstream over a longer reach.

Natural fluctuations in stream discharge in many cases will be sufficient to flush sediment downstream after a temporary disturbance ceases. For example, heavy sediment deposits in Valley Creek, Minnesota, decimated trout populations and invertebrate production, but deposited sediment disappeared by about 2 years after sediment input ceased. The stream had been exposed to natural variations in flow during that time; trout and invertebrates had recovered (Waters 1982).

Many specific techniques have been devised to artificially achieve sediment removal by transfer downstream. Several conditions and limitations, however, are required. First, in order for any sediment removal to be successful in stream restoration, the source of anthropogenic sediment must be eliminated. This requirement, then, calls for the implementation of one or both of the first two control principles described above, prevention and interdiction.

Second, flushing sediment by whatever means does not really remove the material but rather only displaces or transfers it downstream. Any such plan should consider the consequences of the translocation of sediment to downstream reaches or other receiving water bodies. Flushing sediment downstream may successfully restore an upstream reach—and perhaps that may be the sole objective—but the stream as a whole, or the ecosystem in an even broader sense, may not be literally restored unless the anthropogenic sediment is physically removed, possibly to the site of its origin.

Three general procedures have emerged as means to restore sedimented stream reaches: (1) temporary high discharges from dams ("flushing flows") that scour sand and silt and flush sediment downstream; (2) small instream devices that alter fish habitat by locally increasing current velocities to scour certain stream locations favorable to fish, such as fish-holding pools, spawning gravels, and invertebrate producing riffles; and (3) mechanical removal of sediment by gravel washing, riffle cleansing, and other filtering processes (Table 10).

Flushing Flows

Temporary high discharges occur naturally in streams and undoubtedly function to establish the natural character of streams with respect to their "nor-

TABLE 10.—Methods to remove excess sediment from streams.[a]

General practice	Method	Remarks
Flushing flows	Controlled release from dam	Determine magnitude, duration, and timing
	Single pulse	Large pulse may be damaging
	Repeated pulses	More effective with less damage
	Estimating required flows: simulating natural flow regime marking gravel particles	Requires record of past flow regime Movement of gravel indicates movement of fines as well
	engineering formulae	Many in literature, but vary in result
	Post-operation evaluation	Longterm assessment of fine sediment, spawning gravel, fish and invertebrate populations
Stream alteration	Devices to increase water velocity and flush sediment	Like flushing flows (on smaller scale), such devices transfer sediment downstream
	Deflectors to narrow channel	Maintain higher current velocity to clean riffles
	Deflector in combination with bank cover	Deflector aids to maintain sediment-free pool
	Small dam and plunge pool (log drop)	Small waterfall or cascade maintains sediment-free plunge pool; dam should not impound water
	Debris removal	Prevents sediment accumulation behind obstacle
	Eliminate backwaters	Contain water flow to main channel, including in high discharge
Gravel cleaning devices (mechanical removal)	Powered vehicle, to disturb and sift gravel, sediment flows downstream	May cause sediment problem downstream
	Machine to disturb streambed and bury sediment beneath clean gravel	Assumes permanent inactivity of buried sediment
	Machine to disturb streambed, collect and remove fine sediment	"Gravel Gertie," most effective device, requires elimination of sediment source for lasting effect
	Hose and water jet	Method little investigated, transfers sediment downstream, may be most useful in combination with temporary sediment traps
Sediment traps	Sediment-trapping devices to accumulate mainly sand for subsequent dredging and removal	Actually interdiction measures, but can serve effectively to remove sediment over a long reach; series more effective than a large single impoundment

(Continued)

TABLE 10.—Continued.

General practice	Method	Remarks
Sediment traps	Temporary impoundment	Useful during short-term operations, road and bridge construction, urban development
	Permanent dams	Most effective with on-going sediment inputs, may cause aggradation upstream from impoundment
	Excavation in streambed	Temporary or permanent, may cause degradation upstream from excavation
	Larger impoundment in natural waterway	Useful for fish production, must be drainable for periodic sediment removal

[a]Based on: White and Brynildson 1967; Shields 1968; Meehan 1971; Hansen 1973; Brusven et al. 1974; Luedtke and Brusven 1976; Mih 1978; Andrew 1981; Bailey 1981; Alexander and Hansen 1983; Hansen et al. 1983; Reiser et al. 1985, 1987; Bates and Johnson 1986; Kondolf et al. 1987; Wesche et al. 1987; Duff et al. 1988; Lyons and Courtney 1990; Gordon et al. 1992; NRC 1992.

mal" sediment dynamics. Such discharge peaks, spates, and floods can often remove excess anthropogenic sediment over a relatively short term, provided that sediment input is not overwhelming and soon ceases.

Under natural conditions, the sediment transport capability of a given stream may be expected to maintain conditions suitable for native biological communities that have developed under existing conditions (Reiser et al. 1985, 1987). Thus, the assumption is that the natural flow regime is optimum (or at least adequate) with respect to maintaining a sediment condition in the stream that is conducive to the maintenance of normal biological conditions (Kondolf et al. 1987).

Stream regulation by dams and reservoirs eliminates natural variations in flow, including high flows. The result of flushing flows may be excessive sediment accumulations and the potential of damaging fishery resources farther downstream (Nelson et al. 1987). The purpose of flushing flows, sometimes termed "artificial floods" (Gordon et al. 1992), might best be to simulate the natural flow regime (including peak flows) of a specific stream.

The earliest attempt at an experimental field operation was apparently that by Eustis and Hillen (1954), who reported the clearing of a sedimented upstream reach of the Colorado River, Colorado, below a dam installed to divert water to another watershed. The authors used the term "controlled reservoir release." The size of release was determined largely on the basis of the channel capacity downstream, increasing discharge from the normal 20 ft^3/s to about 275–300 ft^3/s. The operation was judged partially successful in that it reduced sediment deposits by 60% and virtually removed all sediment from riffle areas. Tennant (1976) briefly mentioned flushing flows within his recommended instream flow requirements in western streams; he suggested a level of 200% of average flow. Beschta et al. (1981) reported on a "controlled reservoir release" in Huntington Creek, Utah, where discharge was increased from 0.4 to 4.9 m^3/s and maintained for a long period. Suspended and bedload sediments increased greatly downstream from the dam but riffle gravels were not moved. Biological effects were not measured, although the authors stated that such evaluation was needed.

Kondolf et al. (1987) suggested that the simplest method of planning flushing flows is to monitor the effects of natural peaks in flow during a pre-impoundment period and to observe sediment entrainment under a range of flows. Other methods included marking (painting) gravel particles to observe their movement under natural or experimental flows. In this procedure, the assumption is made that if the gravel particles of a salmonid spawning area are entrained or moved, then interstitial sediment fines will also be entrained and moved downstream. Many engineering formulae have been developed to estimate flows required for sediment entrainment from parameters of particle size, slope of water surface, and water depth.

Wesche et al. (1987) studied the North Fork of the Little Snake River, Wyoming, after a heavy deposit of sediment had resulted from construction of a water diversion project. They recommended a flushing flow based on bankfull discharge; they then measured the changes in deposited sediment resulting from natural peaks in discharge for comparison. The sediment removal objective was judged to have been only partly attained; flushing was most useful in the steeper gradient reaches.

The major published contributions to flushing flow methodology have been those by Reiser et al. (1985, 1987, 1989a, 1989b). These include a comprehensive review of flushing flow requirements (Reiser et al. 1985) and a revised, condensed version of this review (Reiser et al. 1987). The authors discussed the rationale for flushing flows, reviewed existing methods, emphasized the importance of evaluation, and offered research needs.

Reiser et al. (1985, 1987) pointed out the need for preliminary evaluation of justification for flushing flows, that is, assessment of present downstream sediment conditions and life history requirements of fishes present. The preoperational study should include a plan for timing, generally based upon a historical record of runoff and seasonal fishery needs. The authors also emphasized the two major determinations required: magnitude of the flow and its duration. Finally, they compiled a list of 15 (Reiser et al. 1985) and 13 (Reiser et al. 1987) methods that have been suggested or developed to determine required magnitudes of flows. Of these, five also addressed the factor of timing, and nine addressed duration (ranging from several hours to 7 days). The authors concluded that no standard, state-of-the-art method exists and pointed out that some of the methods listed were untested. Furthermore, they observed that results from two or more methods could vary widely when applied to the same stream conditions.

The most reliable method that Reiser et al. (1985, 1987) listed involved the observation of the movement of marked (painted) substrate particles under experimental flows. If the empirical conditions necessary for this method were not available, the next most reliable method involved calculations based on sediment transport mechanics and published formulae. The authors provided references leading into the physical-hydrological literature on sediment transport dynamics.

Reiser et al. (1985) also pointed out the need for post-operational evaluation of effectiveness, using the same techniques of sediment and biological sampling as in the preliminary evaluation of need. Reiser et al. (1987) included a method of their own development, termed the "incipient motion method," which estimates the flow required for bed mobilization according to grain size and channel slope.

In a subsequent field application, Reiser et al. (1989a) studied the flushing flow needs for the North Fork of the Feather River, a salmonid stream in California. Because salmonid spawning gravel was limited in this stream, their objective was to remove fine sediment without moving the gravel itself. The North Fork of the Feather River was extremely heterogeneous with respect to streambed and gradient profile (with many boulder cascades) so that estimation of flow needs through use of the theory of bed mobilization—based on uniform flow and substrate—was difficult. Their estimate of magnitude was based on field measurements of low-flow conditions and extrapolations to higher flows. Duration was estimated based on reach length, channel gradient, and pool-to-riffle ratio, using published formulae for sediment transport relationships. They emphasized that their results were specific for the studied stream and resource management objectives.

Reiser et al. (1989b) contributed a book chapter that was a major review of flushing flows. They included a discussion on channel response to river regulation and on existing methods to determine required magnitude and duration. They also provided a guide to recommendations for determining the need for flushing flows, required timing, required magnitude, and evaluation of flushing flow effectiveness. Gordon et al. (1992) in their reference text also reviewed and discussed flushing flow needs and methods of determining flow magnitude, duration, and timing.

Great caution should be used in designing flushing flows because of the severe alterations in stream dynamics, however temporary, that can result in channel morphology changes, streambank erosion, and greater sediment loads downstream from the dam. For example, Nelson et al. (1987) discussed flushing flow methodology with respect to the California Trinity River restoration project. The authors concluded that use of flushing flows was questionable, based partly on the estimate that required flows would damage the few remaining spawning riffles below the dam.

Research needs, pointed out mainly by Reiser et al. (1985), included evaluation of existing methods for calculating required flows and development of new methods; expansion of instream flow and physical habitat studies to include sediment transport considerations; assessment of biological effects of flushing flows; and development of improved sampling techniques for evaluation.

Habitat Alteration

Akin to the massive dam releases of flushing flows involving the scouring of whole reaches of streams, is the action of small instream devices. These devices are often included in habitat alteration or "stream improvement" programs usually undertaken by federal, state, or provincial resource management agencies. The intention is to scour sediment from small, local streambed sites that have particular functional relevance to life-history needs of stream fishes, most commonly in trout fisheries.

Excessive sediment in trout streams produce the greatest threat to: fish-holding pools which sand or silt may fill in an aggrading stream; salmonid redds which may act as sediment traps, reducing reproductive success; and invertebrate-producing riffles of gravel and cobbles in which interstitial space may be filled with sediment and large particles embedded. As in flushing flow operations, little attention appears to have been given to effects downstream where the flushed or scoured sediment may redeposit.

The first comprehensive treatment of trout habitat alteration was the early paper by White and Brynildson (1967). They detailed procedures and structures aimed at correcting physical problems in small streams. The modifications were aimed at controlling or removing sediment in sites critical to fish. Their report was also the first to base these applications upon well-established hydraulic and ecological principles, as well as upon the authors' experience in managing Wisconsin trout streams.

The overriding principle of their guidelines was to maintain or increase water velocity and depth in specific streambed sites, thus scouring sediment from fish-holding pools, bank covers, spawning gravel, and riffles. These objectives were chiefly attained by: placing deflectors alternately to narrow the channel and keep riffles clean; maintaining grass turf on streambanks with open canopy for more sunlight; and excluding livestock from channel and streambanks, as well as other measures to prevent streambank erosion.

White and Brynildson (1967) described specific treatments that included the combined effect of deflector and streambank cover, intended to direct currents under the bank to create and maintain sediment-free holding cover (Figure 21). Small dams were recommended; these were often notched at their center to concentrate scouring in a plunge pool and located at the downstream end of a riffle so as to not create an impoundment. Other treatments were intended to avoid conditions that might trap and hold sediment, such as debris in the channel, structures that might trap water behind the streambank in flood conditions, and any kind of backwater that could trap sediment and later release it to the stream or initiate new channels.

A study on a northern Idaho stream provided some early evaluation of structures aimed at sediment removal (Luedtke and Brusven 1976; Luedtke et al. 1976). The authors tested instream devices and practices such as log drops (plunge pools), debris-jam removal, channel diversions, and gabion deflectors—all intended to scour sand deposits with increased currents. Most of their devices were successful in removing sand and consequently facilitating recolonization of improved substrate by benthic insects. The authors cautioned, however, that the use of such structures, although successful, should not be considered a substitute for the more fundamental solution, elimination of the sediment source (in their case, mining operations).

Similar structures and practices were used in Emerald Creek, Idaho, a stream heavily sedimented from garnet sand mining (Brusven et al. 1974). The structures increased water depth, mean substrate particle size, and percent cobbles in the substrate, with consequent beneficial effects on benthic invertebrates.

Many other publications in the fisheries management literature detail construction, deployment, and effectiveness of channel alterations aimed at sediment

A shallow reach of trout stream with severely eroded streambanks and heavily sedimented streambeds, in a previously cultivated watershed was treated with the renewal of a vegetated and buffered riparian zone. The channel was narrowed and banks were sloped back and replanted, resulting in increased water depths and current velocities, and consequent increases in trout density. Beaver Creek (Whitewater Wildlife Management Area), Minnesota.

control in addition to other objectives. Most of these deal with trout streams; only a few treat warmwater fisheries (Krumholz 1981; Nelson 1988; Lyons and Courtney 1990). A profusion of specific structures has been described in published literature, many designed to provide habitat for fish but also, in part, to remove sediment (White 1975; CDFO 1980; Wesche 1985; Gore and Bryant 1988; Gordon et al. 1992; NRC 1992; Seehorn 1992; Hunt 1993). The removal of sediment that originated specifically in mining wastes was discussed by Starnes (1985).

Many publications have emphasized structures aimed at specific sediment problems. Maughan et al. (1978) described small log dams with falls and cascades that scour and maintain sediment-free pools; Hunt (1969, 1978, 1993) emphasized the deflector–bank cover combination to narrow the channel and produce pools (particularly useful in low-gradient, sandy streams); Gore (1985b) described habitat enhancement structures for benthic macroinvertebrates; Starnes (1985) described the use of instream structures to create habitat in relocated stream channels away from erodible surface mine spoils. Lyons and Courtney (1990) emphasized streambank stabilization but also discussed structures such as wing dams and jetties intended to narrow channels, scour sediment, and expose coarser substrates.

A   **B**   **C**

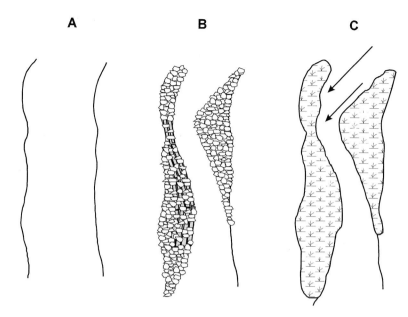

FIGURE 21.—Construction of bank cover for a fish holding pool. (A) Initial wide channel. (B) Construction of bank cover and deflector which narrows the channel and maintains sediment-free pool. (C) Completed structure after vegetation is established (from White and Brynildson 1967).

Apman and Otis (1969) expressed concerns about the possible adverse effects of instream structures. They cautioned about the design of structures that might cause changes in stream morphometry and increase sedimentation. Recently, in a proposed new system of stream classification based on morphological characteristics, Rosgen (1994) pointed out that when fish-habitat structures are installed without regard to possible morphological effects, channel adjustment processes that occur may do more damage than the benefit accrued from the structure. Damages observed by the author in a Colorado stream occurred as the result of aggradation upstream from a gabion-constructed dam and plunge pool; these included lateral channel migration, bank erosion, and excess sediment generation. The author quoted the use of guidelines based on stream morphological conditions by the U.S. Forest Service and others. White and Brynildson (1967) had pointed out that plunge pool dams ("Hewitt dams" and other structures) should always be installed at the lower end of riffles in order to avoid impoundment.

An extensive bibliography on stream habitat alteration is Duff et al. (1988). It contains 1,367 entries for both salmonid and warmwater stream fisheries, including contributions from North America, Europe, and New Zealand.

Gravel Cleaning

Several attempts to develop mechanical means to clean spawning gravels in salmon streams have resulted in at least partial success. The use of such machines appears to be limited to flat, low-gradient areas without sinuosity or boulders.

Consequently, major test applications have been largely limited to artificial spawning channels.

Among the earliest developments was a "riffle sifter" tested in Alaska streams: an amphibious, pneumatic-tracked vehicle that dragged an inverted T-bar through the gravel. From the T-bar, vertically facing nozzles jetted water upward, agitating silt-embedded gravel and raising silty water to the surface. The silt was then collected in a hood riding on the gravel, and the resulting slurry was pumped out of the hood and onto the streambank for disposal (Shields 1968). This machine was tested in three Alaska streams (Slocum Creek, Lovers Cove Creek, and Fish Creek) by Meehan (1971) for its efficiency in sediment reduction in the gravel and also for its effects on the bottom fauna. Reduction of sediment particles smaller than 0.4 mm was 30% in Slocum Creek and 65% in Lovers Cove Creek (Fish Creek was not sampled). Effects on invertebrates included an immediate reduction in the treated area, but abundance increased (greatly, in some cases) after about 3 months, although decreases occurred later.

Andrew (1981) proposed two mechanical methods to clean gravel spawning riffles in incubation channels in the Fraser River system, British Columbia. One of these used a mixture of air and water through subgravel jets, which created upwelling to bring silt to the surface, whence it flowed downstream. The unit was towed over the gravel bed by a track-mounted vehicle ("Gradall"). Field tests in a spawning channel resulted in silt-free depths of gravel from 16 to 20 in. The unit was not effective in natural rivers, however, because boulders interfered with its operation. A second machine, developed for natural rivers, was designed to bury the fines instead of releasing them downstream. It used a vibrating bucket on a Gradall, which dug up a bucketful of gravel, then let it fall in strong currents; the gravel particles fell out and the fines drifted into the hole made by the bucket. The cleaned gravel was then spread on top of the fines. A theoretical calculation of sockeye salmon egg-to-fry survival indicated about a two- or threefold increase in salmon production resulting from the gravel cleaning.

Mih (1978) provided a brief review of various attempts to renovate spawning gravels in salmon streams by reducing fines. These measures included reduction of the source of fines, replacement of spawning beds with new gravel, and use of several devices for mechanical and hydraulic disturbance of gravel to dislodge and remove the fines. Hydraulic disturbance techniques included the "riffle sifter" mentioned above (Shields 1968), to which the author suggested modifications. Mih briefly discussed new developments under way at Washington State University.

The machine developed at the Washington Water Research Center, Washington State University, referred to as "Gravel Gertie" (Bailey 1981; Bates and Johnson 1986), has undergone several modifications and improvements and has been field-tested on various streams in Washington and Idaho (Allen et al. 1981; Mih and Bailey 1981; Bates and Johnson 1986). The cleaning unit is mounted on a self-propelled tractor (otherwise used as a log skidder) that rides on the stream bottom. High-velocity water jets spray downward, at an angle pointing in the direction of movement, to agitate the gravel and move silt upward. An enclosed hood rides on the surface of the gravel, from which the silt slurry is pumped up, concentrated by a centrifugal separator to about 20–40% silt, and then pumped and sprayed out of the channel (Figure 22).

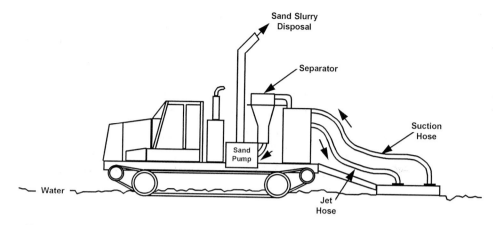

FIGURE 22.—"Gravel Gertie," machine for cleaning sediment from salmonid spawning gravel. The jet hose carries water to disturb gravel and bring silt and sand to surface; the suction hose picks up slurry, which is concentrated in the separator and then pumped out of the stream channel for disposal on land (from Bates and Johnson 1986).

Allen et al. (1981) described preliminary field tests in Idaho and Washington streams. They reported increased survival of fish eggs and fry and increased aquatic insect populations 10–12 months after treatment. However, they also expressed concern for decreased resident trout populations, increased silt deposition downstream, and some problems of the silt slurry re-entering the stream. Bailey (1981) raised additional environmental concerns, pointing out that some native fishes, such as sculpins and minnows that serve as food for salmonids, were eliminated; invertebrate populations experienced some losses in the treated area (some recovery by drift from upstream occurred); a temporary increase in suspended sediment occurred downstream from the machine operations; and a re-suspension of toxic materials, such as mercury used in historic gold mining, was considered a possibility.

Bates and Johnson (1986) presented the most comprehensive report; they described and illustrated the machine ("Gravel Gertie") and reported on field tests in nine Washington streams. Efficiency of removal of fines ranged up to 78% (the mean from nine streams was 65%); aquatic insects, although decreasing during treatment, increased up to 300% in the following year; resident fish numbers decreased only slightly, from 1.08 to 0.90 fish/m^2. The authors cautioned that gravel cleaning would be beneficial only if sediment sources were curtailed. However, if sediment sources were successfully eliminated, natural high seasonal flows might be sufficient to cleanse gravels, and artificial assistance with mechanical gravel cleaning would not be necessary.

The apparent usefulness of gravel-cleaning machines in spawning channels and low-gradient reaches of natural streams suggests that some lighter, similar device might have application in small streams but no such appliance appears to have been developed. The use of gasoline-fueled, engine-driven generators in electrofishing, carried in a small boat towed by a wading crew, suggests that similar operations could be directed to sediment removal. For example, an engine-driven pump might be used with hose and nozzle by a wading crew to flush out fine sediment in small streams. The only reference to such an operation seems to be that of Mundie and Mounce (1978), who used a "fire hose" to flush sand from artificial spawning channels in British Columbia. Such a portable unit, especially in combination with temporary sediment traps (Hansen et al. 1983) that could be frequently dredged during an operation, might be productive in small, highly valued stream fisheries.

SUMMARY

Inorganic fine sediments are naturally present to some extent in all streams. However, in the last half century, excessive sediment of anthropogenic origin has caused enormous damage to streams throughout North America. Most cases that are well-documented by research have been in salmonid streams, particularly in the western United States and Canada. However, most extensive damage to streams has been in the agricultural Midwest and Southeast, where warmwater streams have been severely degraded by excessive sediment. Ironically, it is in these agricultural regions where specific biological effects from sediment has been the least documented.

Quantitatively, sediment has been labeled the most important single pollutant in U.S. streams and rivers. In the latest U.S. Environmental Protection Agency summary of the nation's water quality, siltation tops the list of the foremost 10 pollutants in rivers, half-again higher than the second most important pollutant, nutrients.

Anthropogenic sediment rarely acts alone in its effects on the biological communities in streams. Other factors, such as temperature changes (usually upward), bank slumping and loss of fish habitat, nutrients that cause excessive plant growth, contaminants and chemical pollutants, frequently accompany sedimentation and are often attributable to the same sources or are associated with the same human activities.

SOURCES OF SEDIMENT

Anthropogenic sources of sediment include agriculture (row-crop cultivation, livestock grazing); forestry (timber harvest, logging roads, landslides); mining (spoil piles, tailings dumps, sand and gravel extraction); and urban development (residential, industrial). Road construction for all purposes produces some of the greatest quantities of sediment. Streambank erosion is a natural process in all streams but where it is exacerbated by human activities, destabilized streambanks may deliver great quantities of sediment directly into stream channels.

Agriculture: Row-crop Cultivation

Sedimentation from row-crop cultivation has been overwhelmingly indicted as the major environmental problem in warmwater streams of the Midwest and

Southeast. Reports of the U.S. Environmental Protection Agency list siltation from agriculture as the most important of all river pollutants—more than three times higher than from forestry, mining, or urban development.

Historical reports indicate severe degradation of water quality in many, if not almost all, warmwater streams of the Midwest, with many documented alterations in fish distribution and losses of species. Because such pollution by sediment commenced early along with human settlement in this region, most streams today have few data on pre-cultivation conditions. Surface erosion from poor cultivation practices (up- and downslope plowing, fall plowing, long exposure of disturbed soil), combined with the absence of streambank protection or buffers, has resulted in sediment production in enormous proportions. Particularly indicted is row-crop cultivation on floodplains, where streambanks are severely eroded by floods to deliver sediment directly to stream channels.

Only gross correlations exist between observations of heavily sedimented conditions and apparent loss of fisheries. Specific modes of sediment action on reproductive guilds, feeding guilds, and functional groups related to specific erosional sources have been studied little. Research in these areas is badly needed in order to formulate control and remedial measures.

Agriculture: Livestock Grazing

Sediment sources involving livestock grazing occur mainly in arid, western regions where livestock are attracted to forage and water in the riparian zone. Destabilization of streambanks often results, with large quantities of fine sediments entering the stream. Loss of wildlife habitat in riparian zones, loss of streambank cover for fish, widening and shallowing of the channel, and elevation of water temperatures—all combine to severely degrade both fish and wildlife environments. Salmonid fisheries, largely in the western United States, are affected the most by this source, causing loss of reproductive success in anadromous salmon and steelhead trout fisheries and threatening extirpation of native stocks. Livestock grazing also affects inland trout fisheries in midwestern regions, causing streambank erosion, channel widening, and loss of fish habitat. A large literature exists, both descriptive and experimental, on the effects of overgrazing on sediment production and on stream fisheries. Increasing concern about sedimentation and other damage due to overgrazing has prompted extensive demands for grazing reforms on western rangelands, especially on the vast public lands used for private livestock production.

Forestry

More is known about the potential of sediment from forest management practices than from other sources. Sediment sources are strongly related to the steep hillslopes in western North America and the southern Appalachian Mountains. In such extreme topography (up to 70% gradient) almost any surface disturbance generates potentially damaging sediment. The fisheries affected by excessive sediment in these areas are almost wholly salmonid, including anadromous Pacific salmons and steelhead trout in western regions and inland trout fisheries in the Appalachian region. Potential problems in warmwater fisheries have not been identified or researched relative to potential sediment from forestry practices.

Cutting methods, log skidding and yarding, machinery operation, and site preparation are all important sources of sediment but the overwhelmingly significant source is the logging road. Construction, use, maintenance—even techniques of abandonment—of the logging road have all been strongly indicted.

Two aspects of the logging road problem have been clearly identified: (1) erosion from the roadbed, resulting from overly steep gradients, close proximity to streams, lack of proper drainage, and improper stream crossings; and (2) landslides or mass soil movements that are secondary results of logging road design and placement. Where mass soil movements occur, often unpredictably, the consequent erosion of large areas of the disturbed and exposed soil often contributes the greatest sediment to receiving streams. Sediment, however, is usually not the sole problem generated by landslides; others include debris torrents, destruction of riparian zones and their natural buffering capacity, and clogging of streams with large woody debris, all of which affect stream fisheries.

Several major research programs have been established for experimental study on logging practices: the Alsea Watershed Study in western Oregon; the H. J. Andrews Watershed Study in Oregon; many streams in the granitic batholith in central Idaho, particularly the South Fork Salmon River, an area of highly erodible rock and soils; the Carnation Creek Watershed Project on Vancouver Island, British Columbia; streams of the Olympic Peninsula, Washington; and Coweeta Hydrologic Laboratory, in the southern Appalachian Mountains, North Carolina. All of these intensive, long-term research programs have contributed greatly to identification of sediment sources and the development of control measures.

Mining

Mining operations of several kinds contribute immense quantities of sediment, mainly through erosion of spoil piles, drainage of tailings dumps and ponds, and in sand and gravel extractions. Historically, placer mining—the separation of mineral particles from river-sorted gravels, including streambed deposits—contributed huge quantities of sediment to streams. Although placer mining operations have been curtailed, important gold-mining activity continues in Alaska and northwestern Canada, where inputs of suspended sediment in washwater from ore-treatment operations continue.

Surface mining, both contour strip mining and open-pit, produces large waste piles of overburden. Spoil piles greatly modify whole landscapes and if not regraded and revegetated, can generate enormous quantities of sediment through erosion of exposed spoil surfaces.

Underground mining involves drilling and digging to remove mineral-containing ore and processing to extract the desired mineral, which may be done near the mine site or at processing mills farther away. Processing requires copious amounts of water and results in waste materials or tailings that make up 90% or more of the mined ore. Consequently, tailings are often deposited in ponds, the waters of which may drain away carrying sediment. Unlike most surface-mined spoil piles, processed tailings contain ore particles or solutions that may be toxic to stream organisms.

Sand and gravel mining operations are intimately involved with natural streams, because sand and gravel occur most commonly in alluvial deposits, often in streambeds. Most extractions are in floodplains rather than in the streambed itself, but floodplain operations are closely involved with nearby streams, because gravel and sand washing operations require large quantities of water. Waste water with high concentrations of finer sediment may be returned to the stream. Because sand and gravel extractive mining is the least controlled and regulated— and recorded—of all mining operations, the total extent of these operations is unknown and may be much larger than any other type of mining.

Urban Development

Land-disturbing operations involved in urban construction are varied, including much grading and reshaping of landforms that may expose disturbed soil for extended periods. Specific operations include excavations, soil transport with heavy machinery, drainage during and after construction, channelizing or bridging streams and waterways, temporary storage of soil piles, and road construction. Most published information on municipal sediment sources concerns the large metropolitan complexes around Baltimore, Maryland, and Washington, DC, but some published studies have been conducted in Virginia, North Carolina, and Wisconsin, among others. Sediment contributions from urban areas may be less, on a continental basis, than other major sources. In the smaller, localized areas of municipal construction, however, sediment capable of severe effects on local streams can be generated in extremely high concentrations. Sediment must be treated along with organic pollution, contaminants from industry, and runoff from roads and streets in a comprehensive manner. However, because urban development operations are usually within well-defined boundaries and under potentially strict control from public agencies, sediment control measures may be more effective than in other sources.

Streambank Erosion

Erosion of streambanks is a natural process that results from the tendency of streams to meander. The process may be exacerbated by many human activities that accelerate erosion and generate excessive sediment; the quantities may be large and locally of equal importance to other anthropogenic sources. Two processes are chiefly responsible: (1) entrainment of bank material by high flows, and (2) bank failures that cause slumping of material directly into a stream to be entrained by normal currents. Streambank erosion appears to be a greater problem in warmwater streams of the Midwest, because, historically, settlement and agriculture were attracted to the fertile floodplains of this region. Channelization of streams for any purpose increases streambank erosion potential because the natural tendency of streams is to change straightened channels back to sinuous patterns. Furthermore, straightened channels are shorter with increased gradient and current velocity, which in turn cause further incision and erosion of the streambed.

Miscellaneous Sources

Many other activities can cause erosion of disturbed lands. Road building in any application can have severe erosional effects with consequent sediment generation. Miscellaneous activities that produce sediment include flushing of silted reservoirs, electric transmission and pipeline crossings of streams, bridging and tunneling, habitat alterations in stream fisheries, and any other land- or streambed-disturbing activity. Exposure of disturbed surfaces for long periods extends the duration of erosion potential. Extraordinary storms are a constant threat, exacerbated by past drainage practices, levee construction, and loss of permeable land surface in urban areas. Remedial streamside measures generally have little positive effect on storm-generated sediment production. Not until regional, comprehensive, water-conservation measures are made effective will the disastrous effects of unusual storms be reduced.

BIOLOGICAL EFFECTS OF EXCESS SEDIMENT

Suspended Sediment: Primary Producers

The principal effect of suspended sediment upon primary producers is through turbidity, which reduces light penetration through the water, thus reducing photosynthesis. The ecological effects of sustained, reduced photosynthesis upon higher trophic levels (i.e., invertebrates and fish) is generally unknown. Virtually no experimental research has addressed the effects of sediment-reduced primary production on herbivorous invertebrates.

In some forestry applications, particularly in small headwater streams, light reduction by suspended sediment may be compensated by concurrent canopy removal, which increases sunlight reaching the water surface. The effect of changes in photosynthesis on higher trophic levels presents a difficult problem, because sustained turbidity may have negative effects on invertebrates and fish productivity that may overwhelm effects of photosynthesis variations.

Suspended Sediment: Invertebrates

Much more is known about the effects of suspended sediment on macroinvertebrates. The most common direct effect observed in experiments with fine sediments has been a pronounced increase in downstream drifting. Such increased drift has been attributed primarily to a decrease in light with consequent drift responses similar to behavioral drift in a diel periodicity. Extraordinary drift under prolonged high levels of suspended sediment may deplete benthic invertebrate populations.

Deposited Sediment: Invertebrates

Severe damage to benthic invertebrate populations can be caused by heavy sediment deposits. The factor of "embeddedness"—the fraction of substrate surfaces fixed into surrounding sediment—is a major, measurable parameter that affects the living space of those invertebrates inhabiting the interstitial spaces within the streambed. The affected organisms consist mainly of the insect orders Ephemeroptera, Plecoptera, and Trichoptera, (EPT), which generally are the forms most readily available to foraging fish. Virtually no research has been conducted on the effect of sediment on the meiofauna of streambeds, despite increasing appreciation of the ecological importance of these small organisms to fisheries.

The term "habitat reduction" refers to an increasing level of embeddedness by sediment, resulting in a decrease of invertebrate populations and, consequently, in food available to fish. Complete inundation of gravel and cobble substrate by fine sediment is termed "habitat change," that is, from cobble riffles to homogeneous fine sediment deposits. The consequence of such a habitat change is the alteration of invertebrate forms from EPT to burrowers such as chironomids and oligochaetes, which may be numerous but unavailable to foraging fish. In either case, the fish food resource can be drastically reduced.

Many reports indicate a positive relationship between benthic invertebrate productivity and fish productivity, but direct observational or experimental research on this relationship, as affected by sedimentation, has not been done. Long-term research on the effects of anthropogenic sediment on invertebrate production—and its relationship to fish production—is badly needed, especially in warmwater streams.

Sediment and Fish

Most published information on the effects of sediment in streams relates to fish. Four major categories of this relationship are (1) the direct effects of suspended sediment and turbidity, (2) sediment trapped in salmonid redds and its influence on reproductive success, (3) effects on warmwater fish reproduction, and (4) effects of deposited sediment on fish habitat.

Suspended sediment and fish.—Early studies on the direct effect of suspended sediment on fish generally agreed that fish could withstand concentrations up to 100,000 mg/L at least temporarily but that such high levels were unlikely to be encountered in nature. These early conclusions probably were responsible for delaying the needed intensive investigation into the direct effect of suspended sediment on fish "in nature."

Under conditions where fish have more response options than those available in a laboratory environment, some species exhibit lethal or serious sublethal effects from suspended sediment concentrations far less than 100,000 mg/L. Perhaps more important, fish in nature avoid streams or stream reaches with high suspended sediment levels, creating environments just as devoid of fish as if they had been killed. Intensive, long-term research on this subject is called for, especially in warmwater streams of the Midwest, where so many have been severely degraded, and where almost no experimental research results have been acquired.

Reproductive success of salmonids.—The severe effect of sediment upon developing embryos and sac fry in redds has been intensively investigated. A major problem in many circumstances is that the source of oxygen reaching the redd is in the downwelling water of the stream itself. Suspended sediment carried by stream water enters the redd where velocities are slowed in the interstitial spaces and sediment particles settle. Consequent effects include the coating of eggs and embryos and the filling of interstitial spaces in the redd gravel so completely that the flow of water containing oxygen through the redd is impeded or stopped. The salmonid redd thus functions as an effective "sediment trap," and the entry of oxygen required for embryo survival and development is prevented.

The dependence upon downwelling stream water as the oxygen source appears most common in high-elevation or high-latitude regions, where the stream water source is often cool surface runoff. However, in midwestern regions—with low gradients and warmer runoff—salmonid streams are commonly fed by cool groundwater. Here the oxygen source for the redd is upwelling groundwater that contains no sediment. Consequently, embryo development may continue unimpeded even when stream waters contain high levels of suspended sediment, or when redds contain moderately high quantities of sand that is kept in motion by the upwelling.

A second major problem occurs when sedimentation on the streambed or in upper strata of the redd produces a consolidated armor layer through which emerging sac fry cannot penetrate. Even though embryo development and hatching may be successful within the redd, such entombment of fry attempting to emerge from the redd can result in reproductive failure.

Reproductive success of warmwater fishes.—In contrast to salmonid reproduction, warmwater fish reproduction under various conditions of suspended and deposited sediment is little known. Reproductive behavior of warmwater fish species is more complex and less understood than for salmonid species. Historically, observations of gross correlations between fish species distribution and heavy sedimentation in streams strongly suggested cause and effect but only circumstantial evidence is available. The severe degradation of water quality in streams of the agricultural midwestern and southeastern regions is under increasing scrutiny. Badly needed is experimental and explicit research on the effects of both suspended and deposited sediment, at various levels of intensity, on the specific modes of warmwater fish reproduction.

Deposited sediment and fish habitat.—The effect of deposited sediment, most often sand, on fish-rearing habitat has been studied within two major subject areas: mortality to fry by elimination of interstitial space in riffles of gravel and cobbles, and loss of juvenile-rearing and adult habitat by filling of pools. Salmonid fry, particularly, often require the protection of streambed "roughness" conditions for winter survival. Severe reductions in year-class strength occur when a cohort of salmonid fry faces stream riffles heavily embedded by sediment deposits. Presumably, fry of warmwater species require similar habitat for survival of early life stages, but little research has been accomplished on sediment relationships for these fishes.

Although not as extensively documented, the effect of deposited sediment on juvenile rearing habitat in pools has been similarly convincing. When heavy deposits eliminate pool habitat, reduced growth and loss of populations result.

CONTROL OF SEDIMENT

The control of excess sediment is viewed in three phases: (1) prevention, avoiding erosion or retaining eroded sediment on site; (2) interdiction, capture and retention of sediment between the site of origin and a stream; and (3) restoration, removal of excess sediment from a stream or stream reach.

Prevention

The greatest development of methods to prevent erosion has been in row-crop agriculture. Modern procedures (and some older) include contour plowing, strip cropping, terracing, crop rotation, and recently developed techniques of conservation tillage. Research and development applied to reducing soil losses from row-crop fields have yielded excellent control results when the techniques have been rigidly applied. Field implementation on a broad scale, however, has not been forthcoming. The same techniques developed to reduce soil losses also can be used to reduce sediment pollution in receiving water bodies.

Prevention of sediment generation by livestock overgrazing depends almost solely on fencing to keep livestock from the riparian zone and away from streambanks. The overgrazing problem is most apparent in the arid West but fencing all salmonid streams in arid rangelands appears impracticable. More innovative and effective control techniques have been urged. Similar fencing has been used along inland trout streams in midwestern areas to allow water use by livestock at controlled access points; positive results—more stable streambanks and lower temperatures under streamside canopy—have been achieved.

Many effective techniques in preventing erosion in timber-harvest operations have been developed, particularly for logging roads. Erosion from the roadbed itself can be reduced greatly by surfacing roads with gravel or pavement and by providing cross-drainage to the forest floor with water bars, culverts, and, particularly, the broad-based dip. Seeding roadbeds on sides, on cut and fill slopes, and when abandoned also is effective.

Other factors include road placement, length, and gradient; design of cut and fill slopes; design of stream crossings with culverts and bridges; and harvest methods that require few or no roads, such as overhead skyline yarding and helicopter log transport. The design of roads, especially road cuts, can sometimes be modified to reduce the likelihood of mass soil movements.

The principal way to prevent sediment generation from mining operations is to eliminate erosion of mining wastes—both spoil piles from surface mining and tailings dumps and ponds. The major methods are reshaping spoil piles to an approximation of the original landscape, dewatering of tailings ponds with drainage control, and revegetation. The techniques developed to restore mining-disturbed lands to former terrestrial productivity and aesthetic quality can just as well serve to reduce or eliminate excessive sedimentation of streams with valuable fisheries.

Restoring the physical landscape requires blading and shaping with heavy machinery, reducing gradients, allowing for drainage away from slopes, and replacing topsoil. Revegetation materials include grass seed, transplants, and root stock; grasses, forage, bushes, and trees; and preferably the use of native species.

Irrigation and fertilization may be required. Reclamation of tailings, which often contain toxic mineral particles and solutions, requires special treatment to provide nontoxic soil and suitable conditions for plant growth.

Sand and gravel mining is a very widespread activity that may produce more sediment than any other mining operation. Because sand and gravel are often found in alluvial deposits, the association with streams and rivers is common. Erosion-control measures have not been well developed but some general principles have been suggested: (1) avoid excavations in streambeds entirely; (2) do not connect excavations in floodplains directly to streams; (3) filter water used in washing operations before returning it to streams or underground; and (4) avoid the creation of new stream channels and connecting ponds.

The prevention of sediment generation in urban development depends chiefly on containment of soil and disturbed surface materials on the site. Diverting runoff and sediment from entering a site during construction may forestall having to treat additional sediment; such diversions should direct incoming runoff around the site. Sediment-containment practices aimed at the control of surface runoff include: use of hay, straw, or fiber filters; creation of berms or terraces on contours; installation of retention walls and dams; and construction of small sediment traps or larger settling basins. Measures aimed at reducing the duration of disturbed soil exposure include early revegetation, temporary or permanent sodding, and use of mulches. To promote greater infiltration of runoff water, temporary diversions to level spreaders are also employed. Temporary, unpaved roads can be treated with erosion-control measures similar to those applied on logging roads.

Control of streambank erosion primarily involves grading to reduce bank slope and some kind of bank stabilization structure. Riprap of rock, logs, concrete slabs, and other material may be used on the lower bank to reduce or prevent high stream flows from entraining soil. Sloping is often necessary on both lower and upper banks. Sloping back the upper bank (above high water flows) is intended to prevent bank slumping. The degree of slope should be consistent with the cohesive property of the bank material. Finally, revegetation with grass, bushes, and trees can stabilize surfaces against further erosion from runoff.

Interdiction

Intercepting and retaining sediment between the site of its origin and a receiving stream is second-best to preventing erosion. However, because the total prevention of erosion and soil losses is not yet a reality, interdiction remains an important tool in reducing potential damage in streams. Two general means of interdiction have been developed: (1) buffer strips of vegetation placed or left along streamside to collect and retain sediment as a permanent addition to the riparian zone, and (2) sediment traps placed either on land as temporary structures during construction operations or in the streambed itself as permanent installations which require periodic sediment removal.

Use of buffer strips is one of the oldest means of preventing sediment from reaching streams. Beginning with its application in timber-harvest operations, the buffer strip in its simplest form is a corridor of vegetation left uncut along a stream (i.e., leaving the original riparian zone intact). Buffer strips developed by seeding and planting can also be used effectively in other applications, such as reducing runoff from agricultural areas or urban developments. Fencing the riparian zone in livestock grazing areas is a major application of buffer-strip interdiction.

Much research attention has focused on determining the minimum effective width for a buffer strip. Hillslope and land use are the two most important factors, although watershed area, annual precipitation, and the degree of protection already in place are also important. Buffer strips create stable streamflow, stabilize streambanks, reduce suspended sediment and turbidity, lower summer water temperatures, and filter chemical and organic pollution.

A healthy riparian zone also offers benefit to terrestrial wildlife. Regardless of the research effort and development of different types and sizes of buffer strips, the best application of buffer strip technique is the total avoidance of any surface-disturbing activity in riparian zones and floodplains—cultivation, timber harvest, mining, residential development.

The sediment trap operates on the principal that particles moving in stream currents settle when current velocity is reduced. Sediment traps may be installed in many forms: temporary small structures such as brush barriers used on-land in forestry and urban construction operations; instream excavations and reservoirs to collect bedload sediment that require periodic dredging; and permanent dams and reservoirs that collect sediment and create clearer water downstream. Farm ponds—although usually constructed for livestock watering or fish production—function to trap sediment flowing overland.

Instream sediment traps retain sediment such as active bed load. High dams with large sedimentation capacity, especially when designed with a hypolimnetic water-release feature, create productive and popular tailwater trout fisheries. Although these structures may have a design life of many years, the eventual silting in and loss of reservoir capacity is certain.

On a smaller scale, the instream excavation is most effective in low-gradient streams and is sometimes installed in combination with a dam. The installation is usually designed to trap excess sand in salmonid streams in order to protect spawning beds and invertebrate-producing riffles. The reservoir requires periodic, perhaps frequent, dredging to remove sediment accumulated from both normal and storm flows. The small, instream sediment trap represents very intense management, but it can be extremely effective in small, highly-valued stream fisheries.

Restoration

The term "restoration" in this volume is taken to mean the removal of excess sediment from a stream after the sediment has passed through both the phases of erosion prevention and offsite interdiction. In a program with the goal of complete ecosystem restoration, the removal of excess sediment is only a part—although an essential part—of the overall restorative process.

The success of a removal program depends first on elimination of the sediment source. Most removal measures flush sediment downstream which, more correctly, transfer the sediment, rather than remove it. The effects down-stream are often not known and usually not evaluated.

Three general approaches have been developed for sediment removal: "flushing flows" below large dams; small instream structures to scour and maintain sediment-free sites; and gravel-cleaning machines.

Flushing flows, employed in the operation of large dams and reservoirs, are most commonly used to scour sediment from downstream reaches. In practice, the calculation of appropriate discharge magnitude, timing, and duration has been extremely difficult; a common approach is to attempt to simulate natural fluctuations in discharge. Other problems have been encountered, such as too much scour on existing spawning or food-producing gravel and riffles, increased streambank erosion due to the greater capacity of clearer water to entrain sedi-ment, and the unevaluated transfer of sediment to downstream reaches. Continu-ing research is needed to properly evaluate and further develop the flushing flow technique. Caution is advised in the use of flushing flows because of the changes in stream morphology that may take place downstream and the resulting poten-tial increase in sediment generation from scour, destabilized streambanks, and accelerated incision of the streambed.

Small instream devices are frequently used in the fishery management ap-plication of habitat alteration, usually in trout streams. These devices, often some form of deflectors or other channel-narrowing devices, divert or focus flows to maintain sediment-free spawning gravels, invertebrate-producing riffles, and pools and overhanging streambanks that provide holding cover for fish. The design of many such structures and their use have been profusely published in the past 50 years but relatively little evaluation has been done. Much early work involved placement of logs, stumps, brush, and cribbing without concern for specific habitat needs or hydraulic factors. The more recent and better approach is the judicious application of hydraulic and biological principles to existing stream conditions. As with flushing flows, but on a smaller scale, stream alter-ation devices intended to scour sediment only transfer sediment downstream. Such programs have rarely included the evaluation of possible downstream effects.

Several machines have been designed to sift and remove fine sediments from streambed gravels—mainly to clean salmonid spawning beds. Several designs—usually adapting available tractors or similar machines that power air or water jets through gravel—disturb interstitial deposits of fine sediment, collect the fines in filtering devices, and pump the silt-laden slurry to streamside for disposal. This process actually removes sediment as long as the streamside deposits are placed so as to preclude subsequent return to the stream. This technique represents extremely intensive management and appears to be impracticable in natural rivers that include boulders and varying water depths. However, where the application is on a small but productive spawning reach and the source of sediment has been first eliminated, it can be highly effective.

The installation of small sediment traps, especially in a series and where periodic dredging can be done, can effect removal as well as interdiction. The use of hose, pump, and water jets—although not described in the literature—in combination with a series of frequently dredged sediment traps is suggested as a means of removal of sediment from a small stream—after the original sediment source has been eliminated.

POSTSCRIPT

Enough research now has been completed by biologists of many kinds to know the disastrous results of excess sediment to the major stream biota. Sufficient surveys and summaries have been made by pollution-control agencies to comprehend the enormous quantities of sediment polluting North American streams.

It is also known where the major sediment sources are: agriculture, forestry, mining, urban development.

Recently—and relatively quietly—agricultural engineers, foresters, soil scientists, and many others have developed on-site technologies that markedly reduce soil erosion from sites of human activity, in many cases reducing soil losses almost to zero. The same techniques that retain soils on the land also can prevent the degradation of our streams' biological communities—for public recreational use, aesthetics, and protection of water quality for many purposes.

It is now time to vigorously pursue programs of restoration and planning for future conservation of our streams and rivers, their riparian zones, and the upland landscapes that sustain and nourish them with water, the quality of which we deem worth preserving.

BIBLIOGRAPHY

Adamus, P. R., H. L. Brown, and P. C. Leeper. 1986. Impacts of forest practices on stream ecosystems. Maine Agricultural Experiment Station, Miscellaneous Publication 689, Orono.

Aitkin, W. W. 1936. The relation of soil erosion to stream improvement and fish life. Journal of Forestry 34:1059–1061.

Aldridge, D. W., B. S. Payne, and A. C. Miller. 1987. The effects of intermittent exposure to suspended solids and turbulence on three species of freshwater mussels. Environmental Pollution 45:17–28.

Alexander, G. R., and E. A. Hansen. 1983. Sand sediment in a Michigan trout stream. Part II. Effects of reducing sand bedload on a trout population. North American Journal of Fisheries Management 3:365–372.

Alexander, G. R., and E. A. Hansen. 1986. Sand bed load in a brook trout stream. North American Journal of Fisheries Management 6:9–23.

Allen, J. S., and A. C. Lopinot. 1968. Small lakes and ponds: their construction and care. Illinois Department of Conservation, Fishery Bulletin 3, Springfield.

Allen, K. R. 1951. The Horokiwi Stream, a study of a trout population. New Zealand Marine Department Fisheries Bulletin 10, Wellington.

Allen, R. L., J. E. Seeb, and D. D. King. 1981. A preliminary assessment of field operations with a salmon-spawning gravel cleaning machine. Pages 1–14 *in* WWRC (1981).

Alonso, C. V., and S. T. Combs. 1990. Streambank erosion due to bed degradation—a model concept. Transactions of the American Society of Agricultural Engineers 33: 1239–1248.

American Farmland Trust. 1986. The economics of soil erosion: a handbook for calculating the cost of off-site damage. American Farmland Trust, Washington, DC, and Minnesota Water Conservation Board, St. Paul, Minnesota. (Available from American Farmland Trust, 1920 N Street, N.W., Suite 400, Washington, DC 20036.)

Anderson, D. W. 1983. Factors affecting brown trout reproduction in southeastern Minnesota streams. Minnesota Department of Natural Resources, Investigational Report 376, St. Paul.

Anderson, H. W. 1971. Relative contributions of sediment from source areas, and transport processes. Pages 55–63 *in* Krygier and Hall (1971).

Anderson, H. W., M. D. Hoover, and K. G. Reinhart. 1976. Forests and water: effects of forest management on floods, sedimentation, and water supply. U.S. Forest Service General Technical Report PSW-18.

Anderson, N., and J. Sedell. 1979. Detritus processing by macroinvertebrates in stream ecosystems. Annual Review of Entomology 24:351–377.

Anderson, N. H., J. R. Sedell, L. M. Roberts, and F. J. Triska. 1978. The role of aquatic invertebrates in processing of wood debris in coniferous forest streams. American Midland Naturalist 100:64–82.

Andrew, F. J. 1981. Gravel cleaning to increase salmon production in rivers and spawning channels. Pages 15–31 *in* WWRC (1981).

Anonymous. 1951. Plastic hatching box for stocking trout and salmon. Progressive Fish-Culturist 13:228.

APHA (American Public Health Association), American Water Works Association, and Water Environment Federation. 1992. Standard methods for the examination of water and wastewater, 18th edition. APHA, Washington, DC.

Apmann, R. P., and M. B. Otis. 1965. Sedimentation and stream improvement. New York Fish and Game Journal 12:117–126.

Armour, C. L., D. A. Duff, and W. Elmore. 1991. The effects of livestock grazing on riparian and stream ecosystems. Fisheries 16(1):7–11.

ASSMR (American Society for Surface Mining and Reclamation). 1985. Symposium on the reclamation of lands disturbed by surface mining: a cornerstone for communication and understanding. ASSMR, Owensboro, Kentucky.

Bailey, G. C. 1981. Environmental considerations in gravel rehabilitation operations. Pages 32–37 *in* WWRC (1981).

Baker, R. A., editor. 1980a. Contaminants and sediments, volume 1. Fate and transport, case studies, modeling, toxicity. Ann Arbor Science, Ann Arbor, Michigan.

Baker, R. A., editor. 1980b. Contaminants and sediments, volume 2. Analysis, chemistry, biology. Ann Arbor Science, Ann Arbor, Michigan.

Balon, E. K. 1975. Reproductive guilds of fishes: a proposal and definition. Journal of the Fisheries Research Board of Canada 32:821–864.

Bams, R. A. 1969. Adaptations of sockeye salmon associated with incubation in stream gravels. Pages 71–87 *in* Northcote (1969).

Barnes, J. R., and G. W. Minshall, editors. 1983. Stream ecology: application and testing of general ecological theory. Plenum Press, New York.

Barton, B. A. 1977. Short-term effects of highway construction on the limnology of a small stream in southern Ontario. Freshwater Biology 7:99–108.

Bartsch, A. F. 1959. Settleable solids, turbidity, and light penetration as factors affecting water quality. Pages 118–127 *in* C. M. Tarzwell, editor. Transactions of the second seminar on biological problems in water pollution. U.S. Public Health Service, Robert A. Taft Sanitary Engineering Center, Cincinnati, Ohio.

Bassett, C. E. 1988. Rivers of sand: restoration of fish habitat on the Hiawatha National Forest. Pages 43–48 *in* T. W. Hoekstra and J. Capp, editors. Integrating forest management for fish and wildlife. U.S. Forest Service General Technical Report NC–122.

Bates, J. M. 1962. The impact of impoundment on the mussel fauna of Kentucky Reservoir, Tennessee River. American Midland Naturalist 68:232–236.

Bates, K., and R. Johnson. 1986. A prototype machine for the removal of sediment from gravel streambeds. Washington Department of Fisheries Technical Report 96.

Batie, S. S. 1983. Soil erosion: crisis in America's croplands? The Conservation Foundation, Washington, DC.

Baudo, R., J. P. Giesy, and H. Muntau. 1990. Sediments: chemistry and toxicity of in-place pollutants. Lewis Publishers, Ann Arbor, Michigan.

Baumann, J. S. 1990. Curbing construction site erosion. Wisconsin Natural Resources 14(4):23–28.

Beasley, R. P., J. M. Gregory, and T. R. McCarty. 1984. Erosion and sediment pollution control, 2nd edition. Iowa State University Press, Ames.

Behnke, R. J., and R. F. Raleigh. 1979. Grazing and the riparian zone: impact and management perspectives. Pages 184–189 *in* Johnson and McCormick (1979).

Behnke, R. J., and M. Zarn. 1976. Biology and management of threatened and endangered western trouts. U.S. Forest Service General Technical Report RM-28.

Bell, M. C. 1973. Fisheries handbook of engineering requirements and biological criteria. U.S. Army Corps of Engineers, Fisheries Engineering Research Program, Portland, Oregon.

Bell, D. S., and F. E. Payne. 1993. Mining. Pages 197–207 *in* Bryan and Rutherford (1993).

Benke, A. C., T. C. Van Arsdall, Jr., D. M. Gillespie, and F. K. Parrish. 1984. Invertebrate productivity in a subtropical blackwater river: the importance of habitat and life history. Ecological Monographs 54:25–63.

Bennet, O. L., E. L. Mathias, W. H. Armiger, and J. N. Jones, Jr. 1978. Plant materials and their requirements for growth in humid regions. Pages 285–306 *in* Schaller and Sutton (1978).

Benson, N. G. 1953. The importance of groundwater to trout populations in the Pigeon River, Michigan. Transactions of the North American Wildlife Conference 18:269–281.

Berg, L. 1982. The effect of exposure to short-term pulses of suspended sediment on the behavior of juvenile salmonids. Pages 177–196 *in* Hartman (1982).

Berg, L., and T. G. Northcote. 1985. Changes in territorial, gill-flaring, and feeding behavior in juvenile coho salmon (*Oncorhynchus kisutch*) following short-term pulses of suspended sediment. Canadian Journal of Fisheries and Aquatic Sciences 42:1410–1417.

Berkman, H. E., and C. F. Rabeni. 1987. Effect of siltation on stream fish communities. Environmental Biology of Fishes 18:285–294.

Berry, C. R., Jr. 1979. Impact of sagebrush management on riparian and stream habitat. Pages 192–209 *in* The sagebrush ecosystem: a symposium. Utah State University, College of Natural Resources, Logan.

Beschta, R. L. 1978. Long-term patterns of sediment production following road construction and logging in the Oregon coast range. Water Resources Research 14:1011–1016.

Beschta, R. L. 1982. Comment on 'Stream system evaluation with emphasis on spawning habitat for salmonids' by M. A. Shirazi and W. K. Seim. Water Resources Research 18:1292–1295.

Beschta, R. L., W. L. Jackson, and K. D. Knoop. 1981. Sediment transport during a controlled reservoir release. Water Resources Bulletin 17:635–641.

Bilby, R. E. 1985. Contributions of road surface sediment to a western Washington stream. Forest Science 31:827–838.

Birtwell, I. K., G. F. Hartman, B. Anderson, D. J. McLeay, and J. G. Malick. 1984. A brief investigation of Arctic grayling (*Thymallus arcticus*) and aquatic invertebrates in the Minto Creek drainage, Mayo, Yukon Territory: an area subjected to placer mining. Canadian Technical Report of Fisheries and Aquatic Sciences 1287.

Bisson, P. A., and R. E. Bilby. 1982. Avoidance of suspended sediment by juvenile coho salmon. North American Journal of Fisheries Management 4:371–374.

Bjornn, T. C. 1971. Trout and salmon movements in two Idaho streams as related to temperature, food, streamflow, cover and population density. Transactions of the American Fisheries Society 100:423–438.

Bjornn, T. C., and D. W. Reiser. 1991. Habitat requirements of salmonids in streams. Pages 83–138 *in* Meehan (1991a).

Bjornn, T. C., and seven coauthors. 1974. Sediment in streams and its effects on aquatic life. University of Idaho, Water Resources Research Institute, Research Technical Completion Report Project B-025-IDA, Moscow.

Bjornn, T. C., and six coauthors. 1977. Transport of granitic sediment in streams and its effects on insects and fish. University of Idaho, College of Forestry, Wildlife and Range Sciences, Bulletin 17, Moscow.

Black, P. E. 1991. Watershed hydrology. Prentice Hall, Englewood Cliffs, New Jersey.

Blackmon, B. G., editor. 1985. Proceedings of forestry and water quality: a mid-south symposium. University of Arkansas, Department of Forest Resources, Fayetteville.

Blauch, B. W. 1978. Reclamation of lands disturbed by stone quarries, sand and gravel pits, and borrow pits. Pages 619–628 *in* Schaller and Sutton (1978).

Boccardy, J. A., and W. M. Spaulding, Jr. 1968. Effects of surface mining on fish and wildlife in Appalachia. U.S. Fish and Wildlife Service Resource Publication 65.

Bormann, F. H., G. E. Likens, T. G. Siccama, R. S. Pierce, and J. S Easton. 1974. The export of nutrients and recovery of stable conditions following deforestation at Hubbard Brook. Ecological Monographs 44:255–277.

Braatz, David A. 1993a. Erosion control assessment study: total suspended solids (TSS) in Eastatoe and Little Eastatoe creeks, Pickens County, South Carolina. Research report to Duke Power Company, Huntersville, North Carolina.

Braatz, David A. 1993b. An overview of suspended sediment in the Tellico Creek watershed, in support of a demonstration project for agricultural erosion control. Research report to Duke Power Company, Huntersville, North Carolina.

Branson, B. A., and D. L. Batch. 1972. Effects of strip mining on small-stream fishes in east-central Kentucky. Proceedings of the Biological Society of Washington 84:507–518.

Brigham, A. R., L. B. Suloway, and L. M. Page. 1981. The effects of sedimentation on aquatic life of the Kankakee River. Illinois Department of Energy and Natural Resources Document 81/37, Chicago.

Brown, G. W., and J. T. Krygier. 1971. Clear-cut logging and sediment production in the Oregon coast range. Water Resources Research 7:1189–1198.

Brown, R. J., and J. Mickey, Jr. 1975. Stabilization of erosion on a mountainous road construction project. Pages 87–92 *in* USFS (1975).

Brunskill, G. J., D. M. Rosenberg, N. B. Snow, G. L. Vascotto, and R. Wagemann. 1973. Ecological studies of aquatic systems in the Mackenzie-Porcupine drainages in relation to proposed pipeline and highway developments, volume 1. Northern Pipelines Task Force, Environmental-Social Committee. Northern Oil Development Report 73-40, Winnipeg, Manitoba.

Brusven, M. A., and K. V. Prather. 1974. Influence of stream sediments on distribution of macrobenthos. Journal of the Entomological Society of British Columbia 71:25–32.

Brusven, M. A., and S. T. Rose. 1981. Influence of substrate composition and suspended sediment on insect predation by the torrent sculpin, *Cottus rhotheus*. Canadian Journal of Fisheries and Aquatic Sciences 38:1444–1448.

Brusven, M. A., F. J. Watts, R. Luedtke, and T. L. Kelly. 1974. A model design for physical and biotic rehabilitation of a silted stream. University of Idaho, Water Resources Research Institute, Research Technical Completion Report Project A-032-IDA, Moscow.

Bryan, C. F., and D. A. Rutherford, editors. 1993. Impacts on warmwater streams: guidelines for evaluation. American Fisheries Society, Southern Division, Bethesda, Maryland.

Brynn, D. J., and J. C. Clausen. 1991. Postharvest assessment of Vermont's acceptable silvicultural management practices and water quality impacts. Northern Journal of Applied Forestry 8:140–144.

Bucek, M. F. 1981. Sedimentation ponds and their impact on water quality. Pages 345–354 *in* Symposium on surface mining hydrology, sedimentology and reclamation. University of Kentucky, Lexington.

Burner, C. J. 1951. Characteristics of spawning nests of Columbia River salmon. U.S. Fish and Wildlife Service Fishery Bulletin 52(61).

Burns, J. W. 1970. Spawning bed sedimentation studies in northern California streams. California Fish and Game 56:253–270.

Burns, J. W. 1972. Some effects of logging and associated road construction on northern California streams. Transactions of the American Fisheries Society 101:1–17.

Bustard, D. R., and D. W. Narver. 1975. Preferences of juvenile coho salmon (*Oncorhynchus kisutch*) and cutthroat trout (*Salmo clarki*) relative to simulated alteration of winter habitat. Journal of the Fisheries Research Board of Canada 32:681–687.

Cairns, J., Jr. 1977. Aquatic ecosystem assimilative capacity. Fisheries 2(2):5–7, 24.

Campbell, I. C., and T. J. Doag. 1989. Impact of timber harvesting and production on streams: a review. Australian Journal of Marine and Freshwater Research 40:519–539.

Carr, W. W. 1980. A handbook for forest roadside surface erosion control. British Columbia Ministry of Forests, Land Management Report 4, Victoria.

CDFO (Canadian Department of Fisheries and Oceans). 1980. Stream enhancement guide. CDFO, and British Columbia Ministry of Environment, Vancouver.

Cederholm, C. J., and L. C. Lestelle. 1974. Observations on the effects of landslide siltation on salmon and trout resources of the Clearwater River, Jefferson County, Washington, 1972–73. University of Washington, Fisheries Research Institute Report FRI-UW-7404, Seattle.

Cederholm, C. J., and L. M. Reid. 1987. Impact of forest management on coho salmon (*Oncorhynchus kisutch*) populations of the Clearwater River, Washington: a project summary. Pages 373–398 *in* Salo and Cundy (1987).

Cederholm, C. J., L. M. Reid, and E. O. Salo. 1981. Cumulative effects of logging road sediment on salmonid populations in the Clearwater River, Jefferson County, Washington. Pages 38–74 *in* WWRC (1981).

Cederholm, C. J., and E. O. Salo. 1979. The effects of logging road landslide siltation on the salmon and trout spawning gravels of Stequaleho Creek and the Clearwater River basin, Jefferson County, Washington, 1971–1978. University of Washington, Fisheries Research Institute Report FRI-UW-7915, Seattle.

Cerretani, D. G. 1991. Sampling results of the underground crossing of the Black River. New York State Electric and Gas Corporation, R&D Project 150.51.60, Albany.

Chapman, D. W. 1962. Effects of logging upon fish resources of the West Coast. Journal of Forestry 60:533–537.

Chapman, D. W. 1979. Summarization of sessions. Pages 75–76 *in* Cope (1979).

Chapman, D. W. 1988. Critical review of variables used to define effects of fines in redds of large salmonids. Transactions of the American Fisheries Society 117:1–21.

Chapman, D. W., and T. C. Bjornn. 1969. Distribution of salmonids in streams, with special reference to food and feeding. Pages 153–176 *in* Northcote (1969).

Chapman, D. W., and K. P. McLeod. 1987. Development of criteria for fine sediment in the northern Rockies ecoregion. U.S. Environmental Protection Agency, EPA Report 910/9-87-162, Washington, DC.

Chen, C-N. 1975. Design of sediment retention basins. Pages 285–293 *in* C. T. Haan, editor. National symposium on urban hydrology and sediment control. University of Kentucky, Lexington.

Chisholm J. L., and S. C. Downs. 1978. Stress and recovery of aquatic organisms as related to highway construction along Turtle Creek, Boone County, West Virginia. U.S. Geological Survey Water Supply Paper 2055, Washington, DC.

Christensen, L. A., and P. E. Norris. 1983. A comparison of tillage systems for reducing soil erosion and water pollution. U.S. Economic Research Service, Agricultural Economic Report 499, Washington, DC.

Chutter, F. M. 1969. The effects of silt and sand on the invertebrate fauna of streams and rivers. Hydrobiologia 34:57–76.

Ciborowski, J. J. H., P. J. Pointing, and L. D. Corkum. 1977. The effect of current velocity and sediment on the drift of the mayfly *Ephemerella subvaria* McDunnough. Freshwater Biology 7:567–572.

Claire, E. W., and R. L. Storch. 1983. Streamside management and livestock grazing in the Blue Mountains of Oregon: a case study. Pages 111–128 *in* Menke (1983).

Clark, E. H., II. 1987. Soil erosion: off-site environmental effects. Pages 59–89 *in* Harlin and Berardi (1987).

Clark, E. H., II, J. A. Haverkamp, and W. Chapman. 1985. Eroding soils: the off-farm impacts. The Conservation Foundation, Washington, DC.

Clary, W. P., E. D. McArthur, D. Bedunah, and C. L. Wambolt, editors. 1992. Proceedings, symposium on ecology and management of riparian shrub communities. U.S. Forest Service General Technical Report INT-289.

Cleary, R. E. 1956. Observations of factors affecting smallmouth bass production in Iowa. Journal of Wildlife Management 20:353–359.

Cline, L. D., R. A. Short, and J. V. Ward. 1982. The influence of highway construction on the macroinvertebrates and epilithic algae of a high mountain stream. Hydrobiologia 96:149–159.

Cline, L. D., R. A. Short, J. V. Ward, C. A. Carlson, and H. L. Gary. 1983. Effects of highway construction on water quality and biota in an adjacent Colorado mountain stream. U.S. Forest Service Research Note RM-429.

Cloern, J. E. 1976. The survival of coho salmon (*Oncorhynchus kisutch*) eggs in two Wisconsin tributaries of Lake Michigan. American Midland Naturalist 96:451–461.

Coats, R., L. Collins, J. Florsheim, and D. Kaufman. 1985. Channel change, sediment transport, and fish habitat in a coastal stream: effects of an extreme event. Environmental Management 9:35–48.

Cobb, D. G., and J. F. Flannagan. 1990. Trichoptera and substrate stability in the Ochre River, Manitoba. Hydrobiologia 206:29–38.

Cook, K. 1982. Soil loss: a question of values. Journal of Soil and Water Conservation 37:89–92.

Cooper, A. C. 1965. The effect of transported stream sediments on the survival of sockeye and pink salmon eggs and alevin. International Pacific Salmon Fisheries Commission Bulletin 18.

Cooper, C. M. 1987. Benthos in Bear Creek, Mississippi: effects of habitat variation and agricultural sediments. Journal of Freshwater Ecology 4:101–113.

Cope, O. B., editor. 1979. Proceedings of the forum—grazing riparian/stream ecosystems. Trout Unlimited, Vienna, Virginia.

Cordone, A. J., and D. W. Kelley. 1961. The influences of inorganic sediment on the aquatic life of streams. California Fish and Game 47:189–228.

Cordone, A. J., and S. Pennoyer. 1960. Notes on silt pollution in the Truckee River Drainage, Nevada and Placer counties. California Department of Fish and Game, Inland Fisheries Administrative Report 60-14, Sacramento.

Corn, P. S., and R. B. Bury. 1989. Logging in western Oregon: responses of headwater habitats and stream amphibians. Forest Ecology and Management 29:39–57.

Costa, J. E. 1975. Effects of agriculture on erosion and sedimentation in the Piedmont Province, Maryland. Geological Society of America Bulletin 86:1281–1286.

Crawford, J. K., and D. R. Lenat. 1989. Effects of land use on the water quality and biota of three streams in the Piedmont Province of North Carolina. U.S. Geological Survey, Water Resources Investigations Report 89-4007, Washington, DC.

Crouse, M. R., C. A. Callahan, K. W. Malueg, and S. E. Dominguez. 1981. Effects of fine sediments on growth of juvenile coho salmon in laboratory streams. Transactions of the American Fisheries Society 110:281–286.

Culp, J. M., and R. W. Davies. 1983. An assessment of the effects of streambank clearcutting on macroinvertebrate communities in a managed watershed. Canadian Technical Report of Fisheries and Aquatic Sciences 1208.

Culp, J. M., S. J. Walde, and R. W. Davies. 1983. Relative importance of substrate particle size and detritus to stream benthic macroinvertebrate microdistribution. Canadian Journal of Fisheries and Aquatic Sciences 40:1568–1574.

Culp, J. M., F. J. Wrona, and R. W. Davies. 1986. Response of stream benthos and drift to fine sediment deposition versus transport. Canadian Journal of Zoology 64:1345–1351.

Cummins, A. B., and I. A. Given, editors. 1973a. SME mining engineering handbook, volume 1. Society of Mining Engineers of The American Institute of Mining, Metallurgical, and Petroleum Engineers, Inc., New York.

Cummins, A. B., and I. A. Given, editors. 1973b. SME mining engineering handbook, volume 2. Society of Mining Engineers of The American Institute of Mining, Metallurgical, and Petroleum Engineers, Inc., New York.

Cummins, K. W. 1962. An evaluation of some techniques for the collection and analysis of benthic samples with special emphasis on lotic waters. American Midland Naturalist 67:477–504.

Cummins, K. W. 1964. Factors limiting the microdistribution of larvae of the caddisflies

Pycnopsyche lipida (Hagen) and *Pycnopsyche guttifer* (Walker) in a Michigan stream (Trichoptera:Limnephilidae). Ecological Monographs 34:271–295.

Cummins, K. W. 1966. A review of stream ecology with special emphasis on organism-substrate relationships. Pages 2–51 *in* Cummins et al. (1966).

Cummins, K. W. 1974. Structure and function of stream ecosystems. BioScience 24:631–641.

Cummins, K. W., and M. J. Klug. 1979. Feeding ecology of stream invertebrates. Annual Review of Ecology and Systematics 10:147–172.

Cummins, K. W., and G. H. Lauff. 1969. The influence of substrate particle size on the microdistribution of stream macrobenthos. Hydrobiologia 34:145–181.

Cummins, K. W., C. A. Tryon, Jr., and R. T. Hartman. 1966. Organism-substrate relationships in streams. University of Pittsburgh, Pymatuning Laboratory of Ecology, Special Publication 4, Pittsburgh, Pennsylvania.

Dahlem, E. A. 1979. The Mahogany Creek watershed—with and without grazing. Pages 31–34 *in* Cope (1979).

Dance, K. W., and H. B. N. Hynes. 1980. Some effects of agricultural land use on stream insect communities. Environmental Pollution Series A 22:19–28.

Dechant, T., and J. L. West. 1987. Characteristics and success of wild brown trout redds in 2 western North Carolina streams. Proceedings of the Annual Conference Southeastern Association of Fish and Wildlife Agencies 39(1985):86–96.

Delong, M. D., and M. A. Brusven. 1991. Classification and spatial mapping of riparian habitat with applications toward management of streams impacted by nonpoint source pollution. Environmental Management 15:565–571.

DeVore, P. W., L. T. Brooke, and W. A. Swenson. 1980. The effects of red clay turbidity and sedimentation on aquatic life in the Nemadji River system. Pages 131–265 *in* S. C. Andrews, R. G. Christensen, and C. D. Wilson, editors. Impact of nonpoint pollution control on western Lake Superior. U.S. Environmental Protection Agency, EPA Report 905/9-79-002-B, Washington, DC.

DeWalt, R. E., and J. H. Olive. 1988. Effects of eroding glacial silt on the benthic insects of Silver Creek, Portage County, Ohio. Ohio Journal of Science 88:154–159.

Dieter, C. D. 1990. Causes and effects of water turbidity: a selected annotated bibliography. South Dakota State University, South Dakota Cooperative Wildlife Research Unit, Technical Bulletin 5, Brookings.

Dill, L. M., and T. G. Northcote. 1970. Effects of gravel size, egg depth, and egg density on intragravel movement and emergence of coho salmon (*Oncorhynchus kisutch*) alevins. Journal of the Fisheries Research Board of Canada 27:1191–1199.

Doeg, T. J., and G. A. Milledge. 1991. Effect of experimentally increasing concentrations of suspended sediment on macroinvertebrate drift. Australian Journal of Marine and Freshwater Research 42:519–526.

Douglass, J. E., and O. C. Goodwin. 1980. Runoff and soil erosion from forest site preparation practices. Pages 50–74 *in* Forestry and water quality: what course in the '80s? U.S. Water Pollution Control Federation, Washington, DC.

Douglass, J. E., and L. W. Swift, Jr. 1977. Forest service studies of soil and nutrient losses caused by roads, logging, mechanical site preparation, and prescribed burning in the Southeast. Pages 488–503 *in* D. L. Correll, editor. Watershed research in eastern North America, volume II. Smithsonian Institution, Chesapeake Bay Center for Environmental Studies, Edgewater, Maryland.

Downing, W. L., editor. 1980. Proceedings, National conference on urban erosion and sediment control: institutions & technology. U.S. Environmental Protection Agency, EPA Report 905/9-80-002, Washington, DC.

Duchrow, R. M. 1982. Effects of barite tailings on benthos and turbidity of two Ozark streams. Transactions of the Missouri Academy of Science 16:55–65.

Duda, A. M., D. R. Lenat, and D. Penrose. 1979. Water quality degradation in urban streams of the southeast: will non-point source controls make any difference? Pages

151–159 *in* International symposium on urban storm runoff. University of Kentucky, Lexington.

Duff, D. A. 1983. Livestock grazing impacts on aquatic habitat in Big Creek, Utah. Pages 129–142 *in* Menke (1983).

Duff, D. A., N. Banks, E. Sparks, W. E. Stone, and R. J. Poehlmann. 1988. Indexed bibliography on stream habitat improvement (4th revision). U.S. Forest Service, Intermountain Region, Ogden, Utah.

Duncan, S. H., and J. W. Ward. 1985. The influence of watershed geology and forest roads on the composition of salmon spawning gravel. Northwest Science 59:204–212.

Duncan, W. F. A., and M. A. Brusven. 1985. Benthic macroinvertebrates in logged and unlogged low-order southeast Alaskan streams. Freshwater Invertebrate Biology 4:125–132.

Earhart, H. G. 1984. Monitoring total suspended solids by using nephelometry. Environmental Management 8:81–86.

EIFAC (European Inland Fisheries Advisory Commission). 1965. Water quality criteria for European freshwater fish. Report on finely divided solids and inland fisheries. International Journal of Air and Water Pollution 9:151–168.

Eldridge, E. F., and J. N. Wilson, editors. 1959. Proceedings of the fifth symposium— Pacific Northwest on siltation—its sources and effects on the aquatic environment. U.S. Public Health Service, Water Supply and Water Pollution Control Program, Portland, Oregon.

Elliott, J. M. 1989. The critical-period concept for juvenile survival and its relevance for population regulation in young sea trout, *Salmo trutta*. Journal of Fish Biology 35:91–98.

Ellis, M. M. 1931. A survey of conditions affecting fisheries in the Upper Mississippi River. U.S. Bureau of Fisheries, Fisheries Circular 5, Washington, DC.

Ellis, M. M. 1936. Erosion silt as a factor in aquatic environments. Ecology 17:29–42.

Ellis, M. M. 1944. Water purity standards for fresh-water fishes. U.S. Fish and Wildlife Service Special Scientific Report 2.

Elmore, W. 1992. Riparian responses to grazing practices. Pages 441–457 *in* Naiman (1992).

Elwood, J. W., J. D. Newbold, R. V. O'Neill, and W. Van Winkle. 1983. Resource spiraling: an operational paradigm for analyzing lotic ecosystems. Pages 3–28 *in* T. D. Fontaine, III, and S. M. Bartell, editors. Dynamics of lotic ecosystems. Ann Arbor Science, Ann Arbor, Michigan.

Elwood, J. W., and T. F. Waters. 1969. Effects of floods on food consumption and production rates of a stream brook trout population. Transactions of the American Fisheries Society 98:253–262.

Emmerich, W. E., and J. R. Cox. 1992. Hydrologic characteristics immediately after seasonal burning on introduced and native grasslands. Journal of Range Management 45:476–479.

England, R. H. 1987. Fisheries management on Georgia national forests. Pages 68–71 *in* J. G. Dickson, and O. E. Maughn, editors. Managing southern forests for wildlife and fish, a proceedings. U.S. Forest Service General Technical Report SO-65.

Erman, D. C., and N. A. Erman. 1984. The response of stream macroinvertebrates to substrate size and heterogeneity. Hydrobiologia 108:75–82.

Erman, D. C., and F. K. Ligon. 1988. Effects of discharge fluctuation and the addition of fine sediment on stream fish and macroinvertebrates below a water-filtration facility. Environmental Management 12:85–97.

Erman, D. C., and D. Mahoney. 1983. Recovery after logging in streams with and without bufferstrips in northern California. University of California, Water Resources Center Contribution 186, Davis.

Erman, D. C., J. D. Newbold, and K. B. Roby. 1977. Evaluation of streamside bufferstrips for protecting aquatic organisms. University of California, Water Resources Center Contribution 165, Davis.

Eschmeyer, R. W. 1954. Erosion—and fishing. Montana Wildlife 4(2):21–23.

Eschner, A. R., and J. Larmoyeux. 1963. Logging and trout: four experimental forest practices and their effect on water quality. Progressive Fish-Culturist 25:59–67.

Eustis, A. B., and R. H. Hillen. 1954. Stream sediment removal by controlled reservoir releases. Progressive Fish-Culturist 16:30–35.

Everest, F. H., R. L. Beschta, J. C. Scrivener, K. V. Koski, J. R. Sedell, and C. J. Cederholm. 1987. Fine sediment and salmonid production: a paradox. Pages 98–142 *in* Salo and Cundy (1987).

Everest, F. H., C. E. McLemore, and J. F. Ward. 1980. An improved tri-tube cryogenic gravel sampler. U.S. Forest Service Research Note PNW-350.

Everest, F. H., and W. R. Meehan. 1981. Forest management and anadromous fish habitat productivity. Transactions of the North American Wildlife and Natural Resources Conference 46:521–530.

Everhart, W. H., and R. M. Duchrow. 1970. Effects of suspended sediment on aquatic environment. U.S. Bureau Of Reclamation and Colorado State University, Fort Collins.

Fajen, O. F. 1981. Warmwater stream management with emphasis on bass streams in Missouri. Pages 252–265 *in* Krumholz (1981).

Fajen, O. F., and J. B. Layzer. 1993. Agricultural practices. Pages 257–270 *in* Bryan and Rutherford (1993).

Farnworth, E. G., and seven coauthors. 1979. Impacts of sediment and nutrients on biota in surface waters of the United States. U.S. Environmental Protection Agency, EPA Report 600/3-79-105, Washington, DC.

Faye, R. E., W. P. Carey, J. K. Stamer, and R. L. Kleckner. 1980. Erosion, sediment discharge, and channel morphology in the upper Chattahoochee River basin, Georgia. U.S. Geological Survey Professional Paper 1107, Washington, DC.

Forshage, A., and N. E. Carter. 1974. Effects of gravel dredging on the Brazos River. Proceedings of the annual conference Southeastern Association of Game and Fish Commissioners 27(1973):695–709.

Fraser, J. M. 1985. Shoal spawning of brook trout, *Salvelinus fontinalis*, in a Precambrian Shield lake. Le Naturaliste Canadien 112:163–174.

Fredriksen, R. L. 1965. Sedimentation after logging road construction in a small western Oregon watershed. Pages 56–59 *in* USDA (1965).

Fredriksen, R. L. 1970. Erosion and sedimentation following road construction and timber harvest on unstable soils in three small western Oregon watersheds. U.S. Forest Service Research Paper PNW-104.

Fremling, C. R. 1960. Biology of a large mayfly, *Hexagenia bilineata* (Say), of the Upper Mississippi River. Iowa Agricultural Experiment Station Research Bulletin 482:842–852.

Furniss, M. J., T. D. Roelofs, and C. S. Yee. 1991. Road construction and maintenance. Pages 297–323 *in* Meehan (1991a).

Gabelhouse, D. W., Jr., R. L. Hager, and H. E. Klaassen. 1982. Producing fish and wildlife from Kansas ponds. Kansas Fish and Game Commission, Pratt.

Gammon, J. R. 1970. The effect of inorganic sediment on stream biota. U.S. Environmental Protection Agency, Water Quality Office, Grant Number 18050 DWC, Washington, DC.

Gangmark, H. A., and R. G. Bakkala. 1960. A comparative study of unstable and stable (artificial channel) spawning streams for incubating king salmon at Mill Creek. California Fish and Game 46:151–164.

Garman, G. C., and J. R. Moring. 1987. Impacts of sedimentation on aquatic systems: a literature review and proposed research agenda. Report prepared for the Maine Land Use Regulation Commission, Committee on Stream Research, Orono.

Garman, G. C., and J. R. Moring. 1991. Initial effects of deforestation on physical characteristics of a boreal river. Hydrobiologia 209:29–37.

Gibbons, D. R., W. R. Meehan, K. V. Koski, and T. R. Merrell, Jr. 1987. History of studies of fisheries and forestry interactions in southeastern Alaska. Pages 297–329 *in* Salo and Cundy (1987).

Gibbons, D. R., and E. O. Salo. 1973. An annotated bibliography of the effects of logging on fish of the western United States and Canada. U.S. Forest Service General Technical Report PNW-10.

Glover, R., M. Augustine, and M. Clar. 1978. Grading and shaping for erosion control and rapid vegetative establishment in humid regions. Pages 271–283 *in* Schaller and Sutton (1978).

Goldes, S. A., H. W. Ferguson, R. D. Moccia, and P. -Y. Daoust. 1988. Histological effects of the inert suspended clay kaolin on the gills of juvenile rainbow trout, *Salmo gairdneri* Richardson. Journal of Fish Diseases 11:23–33.

Goldman, S. J., K. Jackson, and T. A. Bursztynsky. 1986. Erosion and sediment control handbook. McGraw-Hill, New York.

Gordon, N. D., T. A. McMahon, and B. L. Finlayson. 1992. Stream hydrology: an introduction for ecologists. Wiley, New York.

Gore, J. A., editor. 1985a. The restoration of rivers and streams: theories and experience. Butterworth, Boston.

Gore, J. A. 1985b. Mechanisms of colonization and habitat enhancement for benthic macroinvertebrates in restored river channels. Pages 81–101 *in* Gore (1985a).

Gore, J. A., and F. L. Bryant. 1988. River and stream restoration. Pages 23–38 *in* J. Cairns, Jr., editor. Rehabilitating damaged ecosystems, volume I. CRC Press, Boca Raton, Florida.

Gottschalk, L. C., and V. H. Jones. 1955. Valleys and hills, erosion and sedimentation. Pages 135–143 *in* USDA (1955).

Gradall, K. S., and W. A. Swenson. 1982. Responses of brook trout and creek chubs to turbidity. Transactions of the American Fisheries Society 111:392–395.

Graham, A. A. 1990. Siltation of stone-surface periphyton in rivers by clay-size particles from low concentrations in suspension. Hydrobiologia 199:107–115.

Gregory, J. D., and J. L. Stokoe. 1981. Streambank management. Pages 276–281 *in* Krumholz (1981).

Gregory, S., and L. Ashkenas. 1990. Riparian management guide, Willamette National Forest. Oregon State University, Department of Fisheries and Wildlife, Corvallis.

Gregory, S. V., G. A. Lamberti, D. C. Erman, K. V. Koski, M. L. Murphy, and J. R. Sedell. 1987. Influence of forest practices on aquatic production. Pages 233–255 *in* Salo and Cundy (1987).

Griffith, J. S., and D. A. Andrews. 1981. Effects of a small suction dredge on fishes and aquatic invertebrates in Idaho streams. North American Journal of Fisheries Management 1:21–28.

Gurtz, M. E., and J. B. Wallace. 1984. Substrate-mediated response of stream invertebrates to disturbance. Ecology 65:1556–1569.

Gurtz, M. E., J. R. Webster, and J. B. Wallace. 1980. Seston dynamics in southern Appalachian streams: effects of clear-cutting. Canadian Journal of Fisheries and Aquatic Sciences 37:624–631.

Guy, H. P., and G. E. Ferguson. 1970. Stream sediment: an environmental problem. Journal of Soil and Water Conservation 25:217–221.

Hachmöller, B., R. A. Matthews, and D. F. Brakke. 1991. Effects of riparian community structure, sediment size, and water quality on the macroinvertebrate communities in a small, suburban stream. Northwest Science 65:125–132.

Hall, J. D., G. W. Brown, and R. L. Lantz. 1987. The Alsea watershed study: a retrospective. Pages 399–416 *in* Salo and Cundy (1987).

Hall, J. D., and R. L. Lantz. 1969. Effects of logging on the habitat of coho salmon and cutthroat trout in coastal streams. Pages 355–375 *in* Northcote (1969).

Hall, T. J. 1984a. The effects of fine sediment on salmonid spawning gravel and juvenile rearing habitat: a literature review. National Council of the Paper Industry for Air and Stream Improvement, Technical Bulletin 428, New York.

Hall, T. J. 1984b. A laboratory study of the effects of sediments of two different size characteristics on survival of rainbow trout (*Salmo gairdneri*) embryos to fry emergence. National Council of the Paper Industry for Air and Stream Improvement, Technical Bulletin 429, New York.

Hall, T. J. 1986. A laboratory study of the effects of fine sediments on survival of three species of Pacific salmon from eyed egg to fry emergence. National Council of the Paper Industry for Air and Stream Improvement, Technical Bulletin 482, New York.

Hall, T. J., R. K. Haley, J. C. Gislason, and G. G. Ice. 1984. The relationship between fine sediment and macroinvertebrate community characteristics: a literature review and results from NCASI fine sediment studies. National Council of the Paper Industry for Air and Stream Improvement, Technical Bulletin 418, New York.

Hall, T. J., and G. G. Ice. 1981. Forest management practices, sediment load characterization, streambed composition and predicting aquatic organism response: where we are and where we need to go. National Council of the Paper Industry for Air and Stream Improvement, Technical Bulletin 353, New York.

Hansen, E. A. 1968. Stabilizing eroding streambanks in sand drift areas of the Lake States. U.S. Forest Service Research Paper NC-21.

Hansen, E. A. 1971. Sediment in a Michigan trout stream, its source, movement, and some effects on fish habitat. U.S. Forest Service Research Paper NC-59.

Hansen, E. A. 1973. In-channel sedimentation basins—possible tool for trout habitat management. Progressive Fish-Culturist 35:138–142.

Hansen, E. A., G. R. Alexander, and W. H. Dunn. 1983. Sand sediment in a Michigan trout stream, part I. A technique for removing sand bedload from streams. North American Journal of Fisheries Management 3:355–364.

Hansen, W. F. 1991. Hazel pistol erosion plot study on the Siskiyou National Forest in southwest Oregon. Pages 107–112 *in* S. C. Nodvin, and T. A. Waldrop, editors. Fire and environment: ecological and cultural perspectives, proceedings of an international symposium. U.S. Forest Service General Technical Report SE-69.

Harlin, J. M., and G. M. Berardi, editors. 1987. Agricultural soil loss: processes, policies, and prospects. Westview Press, Boulder, Colorado.

Harman, W. N. 1972. Benthic substrates: their effect on fresh-water Mollusca. Ecology 53:271–277.

Harman, W. N. 1974. The effects of reservoir construction and canalization on the mollusks of the upper Delaware watershed. Bulletin of the American Malacological Union Incorporated May:12–14.

Harmon, M. E., and twelve coauthors. 1986. Ecology of coarse woody debris in temperate ecosystems. Advances in Ecological Research 15:133–302.

Harrison, C. W. 1923. Planting eyed salmon and trout eggs. Transactions of the American Fisheries Society 53:191–200.

Harshbarger, T. J., and P. E. Porter. 1979. Survival of brown trout eggs: two planting techniques compared. Progressive Fish-Culturist 41:206–209.

Hartman, G. F., editor. 1982. Proceedings of the Carnation Creek workshop: a ten-year review. Canadian Department of Fisheries and Oceans, Nanaimo, British Columbia.

Hartman, G. F., and J. C. Scrivener. 1990. Impacts of forestry practices on a coastal stream ecosystem, Carnation Creek, British Columbia. Canadian Bulletin of Fisheries and Aquatic Sciences 223.

Hartman, G., J. C. Scrivener, L. B. Holtby, and L. Powell. 1987. Some effects of different streamside treatments on physical conditions and fish population processes in Carnation Creek, a coastal rain forest stream in British Columbia. Pages 330–372 *in* Salo and Cundy (1987).

Hartung, R. E., and J. M. Kress. 1977. Woodlands of the northeast: erosion & sediment control guides. U.S. Soil Conservation Service and U.S. Forest Service, Upper Darby, Pennsylvania.

Harvey, B. C. 1986. Effects of suction gold dredging on fish and invertebrates in two California streams. North American Journal of Fisheries Management 6:401–409.

Harvey, M. D., C. C. Watson, and S. A. Schumm. 1985. Gully erosion. U.S. Bureau of Land Reclamation Technical Note 366.

Harwood, G. 1979. A guide to reclaiming small tailings ponds and dumps. U.S. Forest Service General Technical Report INT-57.

Hassett, J. J., and W. L. Banwart. 1992. Soils and their environment. Prentice Hall, Englewood Cliffs, New Jersey.

Haugen, G. N. 1985. American Fisheries Society position statement: strategies for riparian area management. Pages 21–22 in Johnson et al. (1985).

Haupt, H. F. 1959. A method for controlling sediment from logging roads. U.S. Forest Service, Intermountain Forest and Range Experiment Station Miscellaneous Publication 22, Ogden, Utah.

Hausle, D. A., and D. W. Coble. 1976. Influence of sand in redds on survival and emergence of brook trout (Salvelinus fontinalis). Transactions of the American Fisheries Society 105:57–63.

Heede, B. H., and R. M. King. 1990. State-of-the-art timber harvest in an Arizona mixed conifer forest has minimal effect on overland flow and erosion. Hydrological Sciences 35:623–635.

Heede, B. H., and J. N. Rinne. 1990. Hydrodynamic and fluvial morphologic processes: implications for fisheries management and research. North American Journal of Fisheries Management 10:245–268.

Helgath, S. F. 1975. Trail deterioration in the Selway-Bitterroot Wilderness. U.S. Forest Service Research Note INT-193.

Hendrickson, B. H. 1963. Conservation methods for soils of the southern Piedmont. U.S. Department of Agriculture, Agriculture Information Bulletin 269.

Herb, W. J. 1980. Sediment-trap efficiency of a multiple-purpose impoundment, North Branch Rock Creek basin, Montgomery County, Maryland, 1968–1976. U.S. Geological Survey Water Supply Paper 2071, Washington, DC.

Herbert, D. W. M., J. S. Alabaster, M. C. Dart, and R. Lloyd. 1961. The effect of china-clay wastes on trout streams. International Journal of Air and Water Pollution 5:56–74.

Herbert, D. W. M., and J. C. Merkens. 1961. The effect of suspended mineral solids on the survival of trout. International Journal of Air and Water Pollution 5:46–55.

Herricks, E. E., and L. L. Osborne. 1985. Water quality restoration and protection in streams and rivers. Pages 1–20 in Gore (1985a).

Hesse, L. W., and B. A. Newcomb. 1982. Effects of flushing Spencer Hydro on water quality, fish, and insect fauna in the Niobrara River, Nebraska. North American Journal of Fisheries Management 2:45–52.

Hey, R. D., J. C. Bathurst, and C. R. Thorne, editors. 1982. Gravel-bed rivers: fluvial processes, engineering and management. Wiley, New York.

Hill, R. D. 1975. Mining impacts on trout habitat. Pages 47–57 in Proceedings of a symposium on trout habitat research and management. U.S. Forest Service, Southeastern Forest Experiment Station, Asheville, North Carolina.

Hillman, T. W., J. S. Griffith, and W. S. Platts. 1987. Summer and winter habitat selection by juvenile chinook salmon in a highly sedimented Idaho stream. Transactions of the American Fisheries Society 116:185–195.

Hobbs, D. F. 1937. Natural reproduction of quinnat salmon, brown and rainbow trout in certain New Zealand waters. New Zealand Marine Department Fisheries Bulletin 6, Wellington.

Hogan, D. L. 1986. Channel morphology of unlogged, logged, and debris torrented streams in the Queen Charlotte Islands. British Columbia Ministry of Forests and Lands, Land Management Report 49, Victoria.

Horner, R. R., E. B. Welch, M. R. Seeley, and J. M. Jacoby. 1990. Responses of periphyton to changes in current velocity, suspended sediment and phosphorus concentration. Freshwater Biology 24:215–232.

Hubert, W. A., R. P. Lanka, T. A. Wesche, and F. Stabler. 1985. Grazing management influences on two brook trout streams in Wyoming. Pages 290–294 *in* Johnson et al. (1985).

Huehner, M. K. 1987. Field and laboratory determination of substrate preferences of unionid mussels. Ohio Journal of Science 87:29–32.

Hunt, R. L. 1969. Effects of habitat alteration on production, standing crops and yield of brook trout in Lawrence Creek, Wisconsin. Pages 281–312 *in* Northcote (1969).

Hunt, R. L. 1978. Instream enhancement of trout habitat. Pages 19–27 *in* K. Hashagen, editor. Proceedings, a national symposium on wild trout management. California Trout Incorporated, San Francisco.

Hunt, R. L. 1988. Management of riparian zones and stream channels to benefit fisheries. Pages 54–58 *in* T. W. Hoekstra and J. Capp, editors. Integrating forest management for wildlife and fish. U.S. Forest Service General Technical Report NC-122.

Hunt, R. L. 1993. Trout stream therapy. University of Wisconsin Press, Madison.

Hynes, H. B. N. 1966. The biology of polluted waters. Liverpool University Press, Liverpool, UK.

Hynes, H. B. N. 1970. The ecology of running waters. University of Toronto Press, Toronto.

Hynes, H. B. N. 1973. The effects of sediment on the biota in running water. Pages 653–663 *in* Fluvial processes and sedimentation. Canadian Department of the Environment, Ottawa.

Iwamoto, R. N., E. O. Salo, M. A. Madej, and R. L. McComas. 1978. Sediment and water quality: a review of the literature including a suggested approach for water quality criteria. U.S. Environmental Protection Agency, EPA Report 910/9-78-048, Washington, DC.

James, L. A. 1991. Incision and morphologic evolution of an alluvial channel recovering from hydraulic mining sediment. Geological Society of America Bulletin 103:723–736.

Johnson, K. L. 1992. Management for water quality on rangelands through best management practices: the Idaho approach. Pages 415–441 *in* Naiman (1992).

Johnson, R. R., and D. A. Jones, editors. 1977. Importance, preservation, and management of riparian habitat: a symposium. U.S. Forest Service General Technical Report RM-43.

Johnson, R. R., and J. F. McCormick, editors. 1979. Strategies for protection and management of floodplain wetlands and other riparian ecosystems. U.S. Forest Service General Technical Report WO-12.

Johnson, R. R., C. D. Ziebell, D. R. Patton, P. F. Ffolliott, and R. H. Hamre, editors. 1985. Riparian ecosystems and their management: reconciling conflicting uses. U.S. Forest Service General Technical Report RM-120.

Johnson, S. W., J. Heifetz, and K. V. Koski. 1986. Effects of logging on the abundance and seasonal distribution of juvenile steelhead in some southeastern Alaska streams. North American Journal of Fisheries Management 6:532–537.

Jones, R. C., and B. H. Holmes. 1985. Effects of land use practices on water resources in Virginia. Virginia Polytechnic Institute and State University, Water Resources Research Center, Bulletin 144, Blacksburg.

Jordan, D. S. 1889. Report of explorations in Colorado and Utah during the summer of 1889, with an account of the fishes found in each of the river basins examined. U.S. Fish Commission Bulletin 9:1–40.

Judy, R. D., Jr., P. N. Seeley, T. M. Murray, S. C. Svirsky, M. R. Whitworth, and L. S. Ischinger. 1984. 1982 National Fisheries Survey, volume I. Technical report: initial findings. U.S. Fish and Wildlife Service FWS-OBS-84/06.

Kanehl, P., and J. Lyons. 1992. Impacts of in-stream sand and gravel mining on stream habitat and fish communities, including a survey on the Big Rib River, Marathon County, Wisconsin. Wisconsin Department of Natural Resources Research Report 155, Madison.

Karr, J. R., K. D. Fausch, P. L. Angermeier, P. R. Yant, and I. J. Schlosser. 1986. Assessing biological integrity in running waters: a method and its rationale. Illinois Natural History Survey, Special Publication 5, Urbana.

Karr, J. R., and I. J. Schlosser. 1977. Impact of nearstream vegetation and stream morphology on water quality and stream biota. U.S. Environmental Protection Agency, National Environmental Research Center Ecological Research Series EPA-600/3-77-097, Athens, Georgia.

Karr, J. R., L. A. Toth, and D. R. Dudley. 1985. Fish communities of midwestern rivers: a history of degradation. BioScience 35(2):90–95.

Kauffman, J. B., and W. C. Krueger. 1984. Livestock impacts on riparian ecosystems and streamside management implications . . . a review. Journal of Range Management 37:430–438.

Keller, C., L. Anderson, and P. Tappel. 1979. Fish habitat changes in Summit Creek, Idaho, after fencing the riparian area. Pages 46–52 *in* Cope (1979).

Kelley, D. 1962. Sedimentation helps destroy trout streams. Outdoor California 23(3):4, 10–11.

Kidd, W. J., Jr. 1963. Soil erosion control structures on skidtrails. U.S. Forest Service Research Paper INT-1.

Kimble, L. A., and T. A. Wesche. 1975. Relationships between selected physical parameters and benthic community structure in a small mountain stream. University of Wyoming, Water Resources Research Institute, Water Resources Series 55, Laramie.

King, D. L., and R. C. Ball. 1964. The influence of highway construction on a stream. Michigan State University, Agricultural Experiment Station, Research Report 19, East Lansing.

King, D. L., and R. C. Ball. 1967. Comparative energetics of a polluted stream. Limnology and Oceanography 12:27–33.

Kochenderfer, J. N. 1970. Erosion control on logging roads in the Appalachians. U.S. Forest Service Research Paper NE-158.

Kochenderfer, J. N., and P. J. Edwards. 1991. Effectiveness of three streamside management practices in the central Appalachians. Pages 688–700 *in* S. S. Coleman and D. G. Neary, editors. Proceedings of the sixth biennial southern silvicultural research conference. U.S. Forest Service, Southeastern Forest Experiment Station, Asheville, North Carolina.

Kohlhepp, G. W., and R. A. Hellenthal. 1992. The effects of sediment deposition on insect populations and production in a northern Indiana stream. Pages 73–84 *in* T. P. Simon and W. S. Davis, editors. Environmental indicators: measurement and assessment endpoints. U.S. Environmental Protection Agency, EPA Report 905/R-92/003, Washington, DC.

Kondolf, G. M., G. F. Cada, and M. J. Sale. 1987. Assessing flushing-flow requirements for brown trout spawning gravels in steep streams. Water Resources Bulletin 23:927–935.

Koski, K. V. 1966. The survival of coho salmon (*Oncorhynchus kisutch*) from egg deposition to emergence in three Oregon coastal streams. Master's thesis. Oregon State University, Corvallis.

Krug, W. R., and G. L. Goddard. 1986. Effects of urbanization on streamflow, sediment loads, and channel morphology in Pheasant Branch basin near Middleton, Wisconsin. U.S. Geological Survey, Water Resources Investigations Report 85-4068, Washington, DC.

Krumholz, L. A., editor. 1981. The warmwater streams symposium. American Fisheries Society, Southern Division, Bethesda, Maryland.

Krygier, J. T., G. W. Brown, and P. C. Klingeman. 1971. Studies on the effects of watershed practices on streams. U.S. Environmental Protection Agency, Water Pollution Control Research Series, Washington, DC.

Krygier, J. T., and J. D. Hall. 1971. Proceedings of a symposium: forest land uses and stream environment. Oregon State University, Corvallis.

Kunkle, S. H., and G. H. Comer. 1971. Estimating suspended sediment concentrations in streams by turbidity measurements. Journal of Soil and Water Conservation 26:18–20.

Lane, E. W. 1947. Report of the subcommittee on sediment terminology. Transactions of the American Geophysical Union 28:936–938.

Lantz, R. L. 1967. An ecological study of the effects of logging on salmonids. Proceedings of the annual conference Western Association of Game and Fish Commissioners 47:323–335.

Lantz, R. L. 1971. Guidelines for stream protection in logging operations. Report, Oregon State Game Commission, Portland.

LaPerriere, J. D., D. M. Bjerklie, R. C. Simmons, E. Van Nieuwenhuyse, S. M. Wagener, and J. B. Reynolds. 1983. Effects of gold placer mining on interior Alaska stream ecosystems. Pages 12/1–12/34 in J. W. Aldrich, editor. Managing water resources for Alaska's development. University of Alaska, Institute of Water Resources Report IWR-105, Fairbanks.

Larimore, R. W. 1975. Visual and tactile orientation of smallmouth bass fry under flood-water conditions. Pages 323–332 in R. H. Stroud and H. Clepper, editors. Black bass biology and management. Sport Fishing Institute, Washington, DC.

Larimore, R. W., and P. W. Smith. 1963. The fishes of Champaign County, Illinois, as affected by 60 years of stream changes. Illinois Natural History Survey Bulletin 28:299–382.

Larse, R. W. 1971. Prevention and control of erosion and stream sedimentation from forest roads. Pages 76–83 in Krygier and Hall (1971).

Latta, W. C. 1965. Relationship of young-of-the-year trout to mature trout and ground-water. Transactions of the American Fisheries Society 94:32–39.

Lemly, A. D. 1982a. Erosion control at construction sites on red clay soils. Environmental Management 6:343–352.

Lemly, A. D. 1982b. Modification of benthic insect communities in polluted streams: combined effects of sedimentation and nutrient enrichment. Hydrobiologia 87:229–245.

Lenat, D. R. 1984. Agriculture and stream water quality: a biological evaluation of erosion control practices. Environmental Management 8:333–344.

Lenat, D. R. 1988. Water quality assessment of streams using a qualitative collection method for benthic macroinvertebrates. Journal of the North American Benthological Society 7:222–233.

Lenat, D., and K. Eagleson. 1981. Ecological effects of urban runoff on North Carolina streams. North Carolina Department of Natural Resources and Community Development, Biological Series 104, Raleigh.

Lenat, D. R., D. L. Penrose, and K. W. Eagleson. 1979. Biological evaluation of non-point source pollutants in North Carolina streams and rivers. North Carolina Department of Natural Resources and Community Development, Biological Series 102, Raleigh.

Lenat, D. R., D. L. Penrose, and K. W. Eagleson. 1981. Variable effects of sediment addition on stream benthos. Hydrobiologia 79:187–194.

Leopold, L. B. 1994. A view of the river. Harvard University Press, Cambridge, Massachusetts.

Leopold, L. B., M. G. Wolman, and J. P. Miller. 1964. Fluvial processes in geomorphology. Freeman, San Francisco.

Lines, I. L., Jr., J. R. Carlson, and R. A. Corthell. 1979. Repairing flood-damaged streams in the Pacific Northwest. Pages 195–200 in Johnson and McCormick (1979).

Lisle, T. E. 1982. Effects of aggradation and degradation on riffle-pool morphology in natural gravel channels, northwestern California. Water Resources Research 18:1643–1651.

Lisle, T. E., and R. E. Eads. 1991. Methods to measure sedimentation of spawning gravels. U.S. Forest Service Research Note PSW-411.

Lloyd, D. S. 1985. Turbidity in freshwater habitats of Alaska: a review of published and unpublished literature relevant to the use of turbidity as a water quality standard. Alaska Department of Fish and Game, Habitat Division, Report 85-1, Juneau.

Lloyd, D. S. 1987. Turbidity as a water quality standard for salmonid habitats in Alaska. North American Journal of Fisheries Management 7:34–45.

Lloyd, D. S., J. P. Koenings, and J. D. LaPerriere. 1987. Effects of turbidity in fresh waters of Alaska. North American Journal of Fisheries Management 7:18–33.

Lock, M. A., P. N. Wallis, and H. B. N. Hynes. 1977. Colloidal organic carbon in running waters. Oikos 29:1–4.

Lotspeich, F. B., and F. H. Everest. 1981. A new method for reporting and interpreting textural composition of spawning gravel. U.S. Forest Service Research Note PNW-369.

Lotspeich, F. B., and B. Reid. 1980. Tri-tube freeze-core procedure for sampling stream gravels. Progressive Fish-Culturist 42:96–99.

Lowrance, R., R. Leonard, and J. Sheridan. 1985. Managing riparian ecosystems to control nonpoint pollution. Journal of Soil and Water Conservation 40:87–91.

Luedtke, R. J., and M. A. Brusven. 1976. Effects of sand sedimentation on colonization of stream insects. Journal of the Fisheries Research Board of Canada 33:1881–1886.

Luedtke, R. J., M. A. Brusven, and F. J. Watts. 1976. Benthic insect community changes in relation to in-stream alterations of a sediment-polluted stream. Melandaria 23:21–39.

Lum, F. C. H. 1977. Guides for erosion and sediment control in California. U.S. Soil Conservation Service, Miscellaneous Publication, Davis, California.

Lush, D. L., and H. B. N. Hynes. 1973. The formation of particles in freshwater leachates of dead leaves. Limnology and Oceanography 18:968–977.

Lyford, J. H., Jr., and S. V. Gregory. 1975. The dynamics and structure of periphyton communities in three Cascade Mountain streams. Proceedings of the International Association of Theoretical and Applied Limnology 19:1610–1616.

Lyons, J., and C. C. Courtney. 1990. A review of fisheries habitat improvement projects in warmwater streams, with recommendations for Wisconsin. Wisconsin Department of Natural Resources Technical Bulletin 169.

MacCrimmon, H. R., and B. L. Gots. 1986. Laboratory observations on emergent patterns of juvenile Atlantic salmon, *Salmo salar*, relative to sediment loadings of test substrate. Canadian Journal of Zoology 64:1331–1336.

Macdonald, J. S., G. Miller, R. A. Stewart. 1988. The effects of logging, other forest industries and forest management practices on fish: an initial bibliography. Canadian Technical Report of Fisheries and Aquatic Sciences 1622.

MacFarlane, M. B., and T. F. Waters. 1982. Annual production by caddisflies and mayflies in a western Minnesota plains stream. Canadian Journal of Fisheries and Aquatic Sciences 39:1628–1635.

Mackay, R. J., and T. F. Waters. 1986. Effects of small impoundments on hydropsychid caddisfly production in Valley Creek, Minnesota. Ecology 67:1680–1686.

Maidment, D. R., editor. 1993. Handbook of hydrology. McGraw-Hill, New York.

Maine Forest Service. 1991. Erosion & sediment control handbook for Maine timber harvesting operations, best management practices. Maine Forest Service, Augusta.

Marchant, R., and H. B. N. Hynes. 1981. The distribution and production of *Gammarus pseudolimnaeus* (Crustacea: Amphipoda) along a reach of the Credit River, Ontario. Freshwater Biology 11:169–182.

Marcuson, P. E. 1983. Overgrazed streambanks depress fishery production in Rock Creek, Montana. Pages 143–157 in Menke (1983).

Marking, L. L., and T. D. Bills. 1980. Acute effects of silt and sand sedimentation on freshwater mussels. Pages 204–211 in J. L. Rasmussen, editor. Proceedings of the symposium on Upper Mississippi River bivalve mollusks. Upper Mississippi River Conservation Committee, Rock Island, Illinois.

Matthews, G. 1984. Coulee smallmouth: plowed under and suffocated. Wisconsin Natural Resources 8(4):17–20.

Maughan, O. E., K. L. Nelson, and J. J. Ney. 1978. Evaluation of stream improvement practices in southeastern trout streams. Virginia Water Resources Research Center Bulletin 115, Blacksburg.

McClelland, W. T., and M. A. Brusven. 1980. Effects of sedimentation on the behavior and distribution of riffle insects in a laboratory stream. Aquatic Insects 2:161–169.

McCrimmon, H. R. 1954. Stream studies on planted Atlantic salmon. Journal of the Fisheries Research Board of Canada 11:362–403.

McLeay, D. J., I. K. Birtwell, G. F. Hartman, and G. L. Ennis. 1987. Responses of Arctic grayling (*Thymallus arcticus*) to acute and prolonged exposure to Yukon placer mining sediment. Canadian Journal of Fisheries and Aquatic Sciences 44:658–673.

McLeay, D. J., G. L. Ennis, I. K. Birtwell, and G. F. Hartman. 1984. Effects on Arctic grayling (*Thymallus arcticus*) of prolonged exposure to Yukon placer mining sediment: a laboratory study. Canadian Technical Report of Fisheries and Aquatic Sciences 1241.

McLeay, D. J., A. J. Knox, J. G. Malick, I. K. Birtwell, G. Hartman, and G. L. Ennis. 1983. Effects on Arctic grayling (*Thymallus arcticus*) of short-term exposure to Yukon placer mining sediments: laboratory and field studies. Canadian Technical Report of Fisheries and Aquatic Sciences 1171.

McNabb, D. H., and F. J. Swanson. 1990. Effects of fire on soil erosion. Pages 159–176 *in* J. D. Walstad, S. R. Radosevich, and D. V. Sandberg, editors. Natural and prescribed fire in Pacific Northwest forests. Oregon State University Press, Corvallis.

McNeil, W. J., and W. H. Ahnell. 1964. Success of pink salmon spawning relative to size of spawning bed materials. U.S. Fish and Wildlife Service Special Scientific Report Fisheries 469.

Meehan, W. R. 1971. Effects of gravel cleaning on bottom organisms in three southeast Alaska streams. Progressive Fish-Culturist 33:107–111.

Meehan, W. R., editor. 1991a. Influences of forest and rangeland management on salmonid fishes and their habitats. American Fisheries Society Special Publication 19.

Meehan, W. R. 1991b. Introduction and overview. Pages 1–15 *in* Meehan (1991a).

Meehan, W. R., T. R. Merrell, Jr., and T. A. Hanley, editors. 1984. Fish and wildlife relationships in old-growth forests: proceedings of a symposium. American Institute of Fishery Research Biologists, Morehead City, North Carolina.

Meehan, W. R., and W. S. Platts. 1978. Livestock grazing and the aquatic environment. Journal of Soil and Water Conservation 33:274–278.

Meehan, W. R., F. J. Swanson, and J. R. Sedell. 1977. Influences of riparian vegetation on aquatic ecosystems with particular reference to salmonid fishes and their food supply. Pages 137–145 *in* Johnson and Jones (1977).

Megahan, W. F. 1974. Deep-rooted plants for erosion control on granitic road fills in the Idaho batholith. U.S. Forest Service Research Paper INT-161.

Megahan, W. F., and W. J. Kidd. 1972. Effect of logging roads on sediment production rates in the Idaho batholith. U.S. Forest Service Research Paper INT-123.

Megahan, W. F., S. B. Monsen, and M. D. Wilson. 1991. Probability of sediment yields from surface erosion on granitic roadfills in Idaho. Journal of Environmental Quality 20:53–60.

Megahan, W. F., J. P. Potyondy, and K. A. Seyedbagheri. 1992. Best management practices and cumulative effects from sedimentation in the South Fork Salmon River: an Idaho case study. Pages 401–414 *in* Naiman (1992).

Menke, J. W., editor. 1983. Proceedings of the workshop on livestock and wildlife-fisheries relationships in the Great Basin. University of California, Division of Agricultural Sciences, Agricultural Sciences Publication 3301.

Menzel, B. W., J. B. Barnum, and L. M. Antosch. 1984. Ecological alterations of Iowa prairie-agricultural streams. Iowa State Journal of Research 59:5–30.

Meyer, J. L., C. M. Tate, R. T. Edwards, and M. T. Crocker. 1988a. The trophic significance of dissolved organic carbon in streams. Pages 269–278 *in* Swank and Crossley (1988).

Meyer, J. L., and eight coauthors. 1988b. Elemental dynamics in streams. Journal of the North American Benthological Society 7:410–432.

Mih, W. C. 1978. A review of restoration of stream gravel for spawning and rearing of salmon species. Fisheries 3(1):16–18.

Mih, W. C., and G. C. Bailey. 1981. The development of a machine for the restoration of stream gravel for spawning and rearing of salmon. Fisheries 6(6):16–20.

Miller, E. L., R. S. Beasley, and J. C. Covert. 1985. Forest road sediments: production and delivery to streams. Pages 164–176 *in* Blackmon (1985).

Minshall, G. W. 1984. Aquatic insect-substratum relationships. Pages 358–400 *in* Resh and Rosenberg (1984).

Minshall, G. W. 1988. Stream ecosystem theory: a global perspective. Journal of the North American Benthological Society 7:263–288.

Moffett, J. W. 1949. The first four years of king salmon maintenance below Shasta Dam, Sacramento River, California. California Fish and Game 5:77–102.

Moore, E. 1937. The effect of silting on the productivity of waters. Transactions of the North American Wildlife Conference 2:658–661.

Moore, E., E. Janes, F. Kinsinger, K. Pitney, and J. Sainsbury. 1979. Livestock grazing management and water quality protection. U.S. Environmental Protection Agency, EPA Report 910/0-79-67, Washington, DC.

Morgan, R. P., II, V. J. Rasin, Jr., and L. A. Noe. 1983. Sediment effects on eggs and larvae of striped bass and white perch. Transactions of the American Fisheries Society 112:220–224.

Moring, J. R. 1975a. The Alsea watershed study: the effects of logging on the aquatic resources of three headwater streams of the Alsea River, Oregon. Part 2, Changes in environmental conditions. Oregon Department of Fish and Wildlife, Fishery Research Report 9, Corvallis.

Moring, J. R. 1975b. The Alsea watershed study: the effects of logging on the aquatic resources of three headwater streams of the Alsea River, Oregon. Part 3, Discussion and recommendations. Oregon Department of Fish and Wildlife, Fishery Research Report 9, Corvallis.

Moring, J. R. 1982. Decrease in stream gravel permeability after clear-cut logging: an indication of intragravel conditions for developing salmonid eggs and alevins. Hydrobiologia 88:295–298.

Moring, J. R., and G. C. Garman. 1986. The value of riparian zones for fisheries. Pages 81–90 *in* J. A. Bissonette, editor. Is good forestry good wildlife management? University of Maine at Orono Maine Agricultural Experiment Station Miscellaneous Publication 689.

Moring, J. R., G. C. Garman, and D. M. Mullen. 1985. The value of riparian zones for protecting aquatic systems: general concerns and recent studies in Maine. Pages 315–319 *in* Johnson et al. (1985).

Moring, J. R., and R. L. Lantz. 1975. The Alsea watershed study: the effects of logging on the aquatic resources of three headwater streams of the Alsea River, Oregon. Part 1, Biological studies. Oregon Department of Fish and Wildlife, Fishery Research Report 9, Clackamas.

Müller, K. 1974. Stream drift as a chronobiological phenomenon in running water ecosystems. Annual Review of Ecology and Systematics 5:309–323.

Muncy, R. J., G. J. Atchison, R. V. Bulkley, B. W. Menzel, L. G. Perry, and R. C. Summerfelt. 1979. Effects of suspended solids and sediment on reproduction and early life of warmwater fishes: a review. U.S. Environmental Protection Agency, EPA Report 600/3-79-042, Washington, DC.

Mundie, J. H., and D. E. Mounce. 1978. Application of stream ecology to raising salmon smolts in high density. International Association of Theoretical and Applied Limnology Proceedings 20:2013–2018.

Murphy, M. L., and J. D. Hall. 1981. Varied effects of clear-cut logging on predators and their habitat in small streams of the Cascade Mountains, Oregon. Canadian Journal of Fisheries and Aquatic Sciences 38:137–145.

Murphy, M. L., C. P. Hawkins, and N. H. Anderson. 1981. Effects of canopy modification and accumulated sediment on stream communities. Transactions of the American Fisheries Society 110:469–478.

Musser, J. J. 1963. Description of physical environment and of strip-mine operations in parts of Beaver Creek basin, Kentucky. U.S. Geological Survey Professional Paper 427-A, Washington, DC.

Myers, P. C. 1982. Soil erosion and its effects on fish and wildlife resources. Proceedings, convention of the International Association of Fish and Wildlife Agencies 72:36–39.

Naiman, R. J., editor. 1992. Watershed management, balancing sustainability and environmental change. Springer-Verlag, New York.

Nawrot, J. R., A. Woolf, and W. D. Klimstra. 1982. A guide for enhancement of fish and wildlife on abandoned mine lands in the eastern United States. U.S. Fish and Wildlife Service FWS-OBS-80/67.

NCDEHNR (North Carolina Department of Environment, Health and Natural Resources). 1989. Forestry best management practices manual. NCDEHNR, Division of Forest Resources, Raleigh.

Needham, P. R. 1928. A quantitative study of the fish food supply in selected areas. Pages 192–206 in A biological survey of the Oswego River system, supplemental to seventeenth annual report, 1927. New York Conservation Department, Albany.

Nelson, K. L. 1988. Review of warmwater streams habitat improvement methods. North Carolina Wildlife Resources Commission. Federal Aid in Sport Fish Restoration, Project, F-22-13. Final report, Raleigh.

Nelson, K. L. 1993. Instream sand and gravel mining. Pages 189–196 in Bryan and Rutherford (1993).

Nelson, R. L., M. L. McHenry, and W. S. Platts. 1991. Mining. Pages 425–457 in Meehan (1991a).

Nelson, R. W., J. R. Dwyer, and W. E. Greenberg. 1987. Regulated flushing in a gravel-bed river for channel habitat maintenance: a Trinity River fisheries case study. Environmental Management 11:479–493.

Newbold, J. D., D. C. Erman, and K. B. Roby. 1980. Effects of logging on macroinvertebrates in streams with and without buffer strips. Canadian Journal of Fisheries and Aquatic Sciences 37:1076–1085.

Newbury, R. W. 1984. Hydrologic determinants of aquatic insect habitats. Pages 323–357 in Resh and Rosenberg (1984).

Newcomb, T. W., and T. A. Flagg. 1983. Some effects of Mt. St. Helens volcanic ash on juvenile salmon smolts. Marine Fisheries Review 45:8–12.

Newcombe, C. P., and D. D. MacDonald. 1991. Effects of suspended sediments on aquatic ecosystems. North American Journal of Fisheries Management 11:72–82.

Newport, B. D., and J. E. Moyer. 1974. State-of-the-art: sand and gravel industry. U.S. Environmental Protection Agency, EPA Report 660/2-74-066, Washington, DC.

Nielson, R. F., and H. B. Peterson. 1978. Vegetating mine tailings ponds. Pages 645–652 in Schaller and Sutton (1978).

Nobel, E. L., and L. J. Lundeen. 1971. Analysis of rehabilitation treatment alternatives for sediment control. Pages 86–96 in Krygier and Hall (1971).

Noel, D. S., C. W. Martin, and C. A. Federer. 1986. Effects of forest clearcutting in New England on stream macroinvertebrates and periphyton. Environmental Management 10:661–670.

Northcote, T. G., editor. 1969. Symposium on salmon and trout in streams. H. R. MacMillan Lectures in Fisheries, University of British Columbia, Vancouver.

Novinger, G. D., and J. G. Dillard, editors. 1978. New approaches to the management of small impoundments. American Fisheries Society, North Central Division, Special Publication 5, Bethesda, Maryland.

NRC (National Research Council). 1982. Impacts of emerging agricultural trends on fish and wildlife habitat. National Academy Press, Washington, DC.

NRC (National Research Council). 1992. Restoration of aquatic ecosystems: science, technology, and public policy. National Academy Press, Washington, DC.

Nunnally, N. R. 1978. Stream renovation: an alternative to channelization. Environmental Management 2:403–411.

O'Hop, J., and J. B. Wallace. 1983. Invertebrate drift, discharge, and sediment relations in a southern Appalachian headwater stream. Hydrobiologia 98:71–84.

Olsen, L. A., and K. Adams. 1984. Effects of contaminated sediment on fish and wildlife: review and annotated bibliography. U.S. Fish and Wildlife Service FWS-OBS-82/66.

Oregon State University. 1988. Proceedings of the international mountain logging and Pacific Northwest skyline symposium. Oregon State University, Department of Forest Engineering, and International Union of Forestry Research Organizations, Portland.

ORVWSC (Ohio River Valley Water Sanitation Commission). 1956. Aquatic life water quality criteria, second progress report. Sewage and Industrial Wastes 28:678–690.

Oschwald, W. R. 1972. Sediment-water interactions. Journal of Environmental Quality 1:360–366.

Ottens, J., and J. Rudd. 1977. Environmental protection costs in logging road design and construction to prevent increased sedimentation in the Carnation Creek watershed. Centre for Forest Studies, Pacific Forest Research Centre, Victoria, British Columbia.

Overcash, M. R., and J. M. Davidson, editors. 1983. Environmental impact of nonpoint source pollution. Ann Arbor Science, Ann Arbor, Michigan.

Packer, P. E., and E. F. Aldon. 1978. Revegetation techniques for dry regions. Pages 425–450 *in* Schaller and Sutton (1978).

Pain, S. 1987. After the goldrush. Newscientist 115(1574):36–40.

Paone, J., P. Struthers, and W. Johnson. 1978. Extent of disturbed lands and major reclamation problems in the United States. Pages 11–22 *in* Schaller and Sutton (1978).

Paragamian, V. L. 1981. Some habitat characteristics that affect abundance and winter survival of smallmouth bass in the Maquoketa River, Iowa. Pages 45–53 *in* L. A. Krumholz, editor. The warmwater streams symposium. American Fisheries Society, Southern Division, Bethesda, Maryland.

Parker, C. R., and J. R. Voshell, Jr. 1983. Production of filter-feeding Trichoptera in an impounded and a free-flowing river. Canadian Journal of Zoology 61:70–87.

Parmalee, P. W. 1993. Freshwater mussels (Mollusca: Pelecypoda: Unionidae) of Tellico Lake: twelve years after impoundment of the Little Tennessee River. Annals of Carnegie Museum 62:81–93.

Patrick, R. 1976. The importance of monitoring change. Pages 157–190 *in* J. Cairns, Jr., K. L. Dickson, and G. F. Westlake, editors. Biological monitoring of water and effluent quality: a symposium. American Society for Testing and Materials, Philadelphia, Pennsylvania.

Pautzke, C. F. 1938. Studies on the effect of coal washings on steelhead and cutthroat trout. Transactions of the American Fisheries Society 67:232–233.

Pennak, R. W., and E. D. Van Gerpen. 1947. Bottom fauna production and physical nature of the substrate in a northern Colorado trout stream. Ecology 28:42–48.

Penrose, D., D. Lenat, and K. Eagleson. 1980. Biological evaluation of water quality in North Carolina streams and rivers. North Carolina Department of Natural Resources and Community Development, Biological Series 103, Raleigh.

Pentland, E. S. 1930. Controlling factors in the distribution of *Gammarus*. Transactions of the American Fisheries Society 60:89–94.

Peters, J. C. 1965. The effects of stream sedimentation on trout embryo survival. Pages 275–279 *in* C. M. Tarzwell, editor. Biological problems in water pollution. U.S. Public Health Service, Robert A. Taft Sanitary Engineering Center, Cincinnati, Ohio.

Peters, J. C. 1967. Effects on a trout stream of sediment from agricultural practices. Journal of Wildlife Management 31:805–812.

Peterson, A. 1990. The impact of installation and use of electric transmission rights-of-way access roads on trout stream quality. New York State Electric and Gas Corporation, R & D Project 150.53.50, Albany.

Peterson, R. H., and J. L. Metcalfe. 1981. Emergence of Atlantic salmon fry from gravels of varying composition: a laboratory study. Canadian Technical Report of Fisheries and Aquatic Sciences 1020.

Petticord, R. K. 1980. Direct effects of suspended sediments on aquatic organisms. Pages 501–536 *in* Baker (1980a).

Phillips, R. W. 1971. Effects of sediment on the gravel environment and fish production. Pages 64–74 *in* Krygier and Hall (1971).

Phillips, R. W., R. L. Lantz, E. W. Claire, and J. R. Moring. 1975. Some effects of gravel mixtures on emergence of coho salmon and steelhead trout fry. Transactions of the American Fisheries Society 104:461–466.

Platts, W. S. 1968. South Fork Salmon River, Idaho, aquatic habitat survey with evaluation of sediment accruement, movement, and damages. U.S. Forest Service, Intermountain Region, Ogden, Utah.

Platts, W. S. 1970. The effects of logging and road construction on the aquatic habitat of the South Fork Salmon River, Idaho. Proceedings of the annual conference of the Western Association of Game and Fish Commissioners 50:182–185.

Platts, W. S. 1979. Livestock grazing and riparian/stream ecosystems, an overview. Pages 39–45 *in* Cope (1979).

Platts, W. S. 1981a. Effects of sheep grazing on a riparian-stream environment. U.S. Forest Service Research Note INT-307.

Platts, W. S. 1981b. Influence of forest and rangeland management on anadromous fish habitat in western North America: 7. Effects of livestock grazing. U.S. Forest Service General Technical Report PNW-124.

Platts, W. S. 1982. Livestock and riparian fishery interactions: what are the facts? Transactions of the North American Wildlife and Natural Resources Conference 47:507–515.

Platts, W. S. 1991. Livestock grazing. Pages 389–423 *in* Meehan (1991a).

Platts, W. S., K. A. Gebhardt, and W. A. Jackson. 1985. The effects of large storm events on basin-range riparian stream habitats. Pages 30–34 *in* Johnson et al. (1985).

Platts, W. S., and S. B. Martin. 1980. Livestock grazing and logging effects on trout. Pages 34–46 *in* W. King, editor. Wild Trout II. Trout Unlimited and Federation of Fly Fishers, Vienna, Virginia.

Platts, W. S., and W. F. Megahan. 1975. Time trends in riverbed sediment composition in salmon and steelhead spawning areas: South Fork Salmon River, Idaho. Transactions of the North American Wildlife Conference 40:229–239.

Platts, W. S., W. F. Megahan, and G. W. Minshall. 1983. Methods for evaluating stream, riparian, and biotic conditions. U.S. Forest Service General Technical Report INT-138.

Platts, W. S., and R. L. Nelson. 1985a. Impacts of rest-rotation grazing on stream banks in forested watersheds in Idaho. North American Journal of Fisheries Management 5:547–556.

Platts, W. S., and R. L. Nelson. 1985b. Stream habitat and fisheries response to livestock grazing and instream improvement structures, Big Creek, Utah. Journal of Soil and Water Conservation 40:374–379.

Platts, W. S., and V. E. Penton. 1980. A new freezing technique for sampling salmonid redds. U.S. Forest Service Research Paper INT-248.

Platts, W. S., M. A. Shirazi, and D. H. Lewis. 1979. Sediment particle sizes used by salmon for spawning, with methods for evaluation. U.S. Environmental Protection Agency, EPA Report 600/3-79-043, Washington, DC.

Platts, W. S., and thirteen coauthors. 1983. Livestock interactions with fish and their environments. Pages 36–41 *in* Menke (1983).

Platts, W. S., R. J. Torquemada, M. L. McHenry, and C. K. Graham. 1989. Changes in salmon spawning and rearing habitat from increased delivery of fine sediment to the South Fork Salmon River, Idaho. Transactions of the American Fisheries Society 118:274–283.

Quinn, J. M, and C. W. Hickey. 1990. Magnitude of effects of substrate particle size, recent flooding, and catchment development on benthic invertebrates in 88 New Zealand rivers. New Zealand Journal of Marine and Freshwater Research 24:411–427.

Rader, R. B., and J. V. Ward. 1989. Influence of impoundments on mayfly diets, life histories, and production. Journal of the North American Benthological Society 8:64–73.

Randall, C. W., T. J. Grizzard, and R. C. Hoehn. 1978. Impact of urban runoff on water quality in the Occoquan watershed. Virginia Polytechnic Institute and State University, Water Resources Research Center Bulletin 80.

Redding, J. M., C. B. Schreck, and F. H. Everest. 1987. Physiological effects on coho salmon and steelhead of exposure to suspended solids. Transactions of the American Fisheries Society 116:737–744.

Reed, J. R., Jr. 1977. Stream community response to road construction sediments. Virginia Polytechnic Institute and State University, Water Resources Research Center Bulletin 97.

Reeves, G. H., J. D. Hall, T. D. Roelofs, T. L. Hickman, and C. O. Baker. 1991. Rehabilitating and modifying stream habitats. Pages 519–557 in Meehan (1991a).

Reid, L. M., and T. Dunne. 1984. Sediment production from forest road surfaces. Water Resources Research 20:1753–1761.

Reinert, V., and W. P. Oemichen. 1976. Erosion: today's soil, tomorrow's silt. The Minnesota Volunteer March–April:36–43.

Reinhart, K. G., A. R. Eschner, and G. R. Trimble, Jr. 1963. Effect on streamflow of four forest practices in the mountains of West Virginia. U.S. Forest Service Research Paper NE-1.

Reiser, D. W., and T. C. Bjornn. 1979. Influence of forest and rangeland management on anadromous fish habitat in the western United States and Canada: 1. Habitat requirements of anadromous salmonids. U.S. Forest Service General Technical Report PNW-96.

Reiser, D. W., M. P. Ramey, S. Beck, T. R. Lambert, and R. E. Geary. 1989a. Flushing flow recommendations for maintenance of salmonid spawning gravels in a steep, regulated stream. Regulated Rivers Research & Management 3:267–275.

Reiser, D. W., M. P. Ramey, and T. R. Lambert. 1985. Review of flushing flow requirements in regulated streams. Completion report to Pacific Gas and Electric Company, Department of Engineering Research, San Ramon, California.

Reiser, D. W., M. P. Ramey, and T. R. Lambert. 1987. Considerations in assessing flushing flow needs in regulated stream systems. Pages 45–57 in J. F. Craig and J. B. Kemper, editors. Regulated streams: advances in ecology. Plenum Press, New York.

Reiser, D. W., M. P. Ramey, and T. A. Wesche. 1989b. Flushing flows. Pages 91–135 in J. A. Gore and G. W. Petts, editors. Alternatives in regulated river management. CRC Press, Boca Raton, Florida.

Reiser, D. W., and R. G. White. 1988. Effects of two sediment size-classes on survival of steelhead and chinook salmon eggs. North American Journal of Fisheries Management 8:432–437.

Resh, V. H., and D. M. Rosenberg, editors. 1984. The ecology of aquatic insects. Praeger Publishers, New York.

Reynolds, J. B., R. C. Simmons, and A. R. Burkholder. 1989. Effects of placer mining discharge on health and food of Arctic grayling. Water Resources Bulletin 25:625–635.

Rhoads, B. L. 1994. Fluvial geomorphology. Progress in Physical Geography 18:103–123.

Ribaudo, M. O. 1986. Reducing soil erosion: offsite benefits. U.S. Department of Agriculture, Agricultural Economic Report 561, Washington, DC.

Rice, R. M., and J. Lewis. 1991. Estimating erosion risks associated with logging and forest roads in northwestern California. Water Resources Bulletin 27:809–818.

Rice, R. M., J. S. Rothacher, and W. F. Megahan. 1972. Erosional consequences of timber harvest: an appraisal. Pages 321–329 in S. C. Csallany, T. G. McLaughlin, and W. D. Striffler, editors. Watersheds in transition, proceedings of a symposium. American Water Resources Association Publications, Urbana, Illinois.

Rice, R. M., F. B. Tilley, and P. A. Datzman. 1979. A watershed's response to logging and roads: South Fork of Caspar Creek, California, 1967–1976. U.S. Forest Service Research Paper PSW-146.

Richards, K. S. 1982. Rivers, form and process in alluvial channels. Metheun, New York.

Richardson, B. Z., and M. M. Pratt, editors. 1980. Environmental effects of surface mining of minerals other than coal: annotated bibliography and summary report. U.S. Forest Service General Technical Report INT-95.

Ringler, N. H., and J. D. Hall. 1988. Vertical distribution of sediment and organic debris in coho salmon (*Oncorhynchus kisutch*) redds in three small Oregon streams. Canadian Journal of Fisheries and Aquatic Sciences 45:742–747.

Rinne, J. N. 1985. Livestock grazing effects on southwestern streams: a complex research problem. Pages 295–299 *in* Johnson et al. (1985).

Ritchie, J. C. 1972. Sediment, fish, and fish habitat. Journal of Soil and Water Conservation 27:124–125.

Rivier, B., and J. Seguier. 1985. Physical and biological effects of gravel extraction in river beds. Pages 131–146 *in* J. S. Alabaster, editor. Habitat modification and freshwater fisheries. Butterworth, London.

Roberts, R. G. 1987. Stream channel morphology: major fluvial disturbances in logged watersheds on the Queen Charlotte Islands. British Columbia Ministry of Forests and Lands, Land Management Report 48, Victoria.

Robinson, A. R. 1971. A primer on agricultural pollution: sediment. Journal of Soil and Water Conservation 26:61–62.

Roby, K. B., D. C. Erman, and J. D. Newbold. 1977. Biological assessment of timber management activity impacts and buffer strip effectiveness on national forest streams of northern California. U.S. Forest Service, Earth Resources Monograph 1, San Francisco.

Roseboom, D., and K. Russell. 1985. Riparian vegetation reduces stream bank and row crop flood damages. Pages 241–244 *in* Johnson et al. (1985).

Roseboom, D., R. Sauer, and D. Day. 1990. Application of a trout habitat improvement technique to Illinois warmwater streams. *In* The restoration of midwestern stream habitat. American Fisheries Society, North Central Division, Rivers and Streams Technical Committee. (Paper read at the 52nd midwest fish and wildlife conference, December 4–5, 1990, Minneapolis, Minnesota.)

Rosenberg, D. M., and N. B. Snow. 1975a. Ecological studies of aquatic organisms in the Mackenzie and Porcupine river drainages in relation to sedimentation. Department of the Environment, Fisheries and Marine Service, Freshwater Institute, Technical Report 547, Winnipeg, Manitoba.

Rosenberg, D. M., and N. B. Snow. 1975b. A design for environmental impact studies with special reference to sedimentation in aquatic systems of the Mackenzie and Porcupine river drainages. Pages 65–78 *in* Proceedings of the circumpolar conference on northern ecology. National Research Council of Canada, Ottawa.

Rosenberg, D. M., and A. P. Wiens. 1978. Effects of sediment addition on macrobenthic invertebrates in a northern Canadian river. Water Research 12:753–763.

Rosenberg, D. M., and A. P. Wiens. 1980. Responses of Chironomidae (Diptera) to short-term experimental sediment additions in the Harris River, Northwest Territories, Canada. Acta Universitatis Carolinae-Biologica 1978:181–192.

Rosenberg, G. D., and M. T. Henschen. 1986. Sediment particles as a cause of nacre staining in the freshwater mussel, *Amblema plicata* (Say) (Bivalia: Unionidae). Hydrobiologia 135:167–178.

Rosgen, D. L. 1994. A classification of natural rivers. Catena 22(3):169–199.

Rutherford, J. E., and R. J. Mackay. 1986. Patterns of pupal mortality in field populations of *Hydropsyche* and *Cheumatopsyche* (Trichoptera, Hydropsychidae). Freshwater Biology 16:337–350.

Rutherford, R. 1986. Forest harvesting and fish habitat. Maine Agricultural Experiment Station Miscellaneous Publication 689:167–180.

Ryan, P. A. 1991. Environmental effects of sediment on New Zealand streams: a review. New Zealand Journal of Marine and Freshwater Research 25:207–221.

Salo, E. O., and T. W. Cundy, editors. 1987. Streamside management: forestry and fishery interactions. University of Washington Institute of Forest Resources Contribution 57.

Saunders, J. W., and M. W. Smith. 1965. Changes in a stream population of trout associated with increased silt. Journal of the Fisheries Research Board of Canada 22:395–404.

Scannell, P. O. 1988. Effects of elevated sediment levels from placer mining on survival and behavior of immature Arctic grayling. Alaska Cooperative Fishery Research Unit, University of Alaska Fairbanks, Unit Contribution 27, Fairbanks.

Schaller, F. W., and P. Sutton, editors. 1978. Reclamation of drastically disturbed lands. American Society of Agronomy, Crop Science Society of America, and Soil Science Society of America, Madison, Wisconsin.

Schwab, G. O., D. D. Fangmeier, W. J. Elliot, and R. K. Frevert. 1993. Soil and water conservation engineering, 4th edition. Wiley, New York.

Scrivener, J. C., and M. J. Brownlee. 1982. An analysis of the Carnation Creek gravel quality data 1973 to 1981. Pages 154–173 in Hartman (1982).

Scrivener, J. C., and M. J. Brownlee. 1989. Effects of forest harvesting on spawning gravel and incubation survival of chum (Oncorhynchus keta) and coho salmon (Oncorhynchus kisutch) in Carnation Creek, British Columbia. Canadian Journal of Fisheries and Aquatic Sciences 46:681–696.

Seehorn, M. E. 1987. The influence of silvicultural practices on fisheries management: effects and mitigation measures. Pages 54–63 in Managing southern forests for wildlife and fish, a proceedings. U.S. Forest Service General Technical Report SO-65.

Seehorn, M. E. 1992. Stream habitat improvement handbook. U.S. Forest Service Technical Publication R8-TP 16.

Servizi, J. A., and D. W. Martens. 1991. Effect of temperature, season, and fish size on acute lethality of suspended sediments to coho salmon (Oncorhynchus kisutch). Canadian Journal of Fisheries and Aquatic Sciences 48:493–497.

Servizi, J. A., and D. W. Martens. 1992. Sublethal responses of coho salmon (Oncorhynchus kisutch) to suspended sediments. Canadian Journal of Fisheries and Aquatic Sciences 49:1389–1395.

Seyedbagheri, K. S., M. L. McHenry, and W. S. Platts. 1987. An annotated bibliography of the hydrology and fishery studies of the South Fork Salmon River, Idaho. U.S. Forest Service General Technical Report INT-235.

Shapley, S. P., and D. M. Bishop. 1965. Sedimentation in a salmon stream. Journal of the Fisheries Research Board of Canada 22:919–928.

Shapovalov, L., and W. Berrian. 1940. An experiment in hatching silver salmon (Oncorhynchus kisutch) eggs in gravel. Transactions of the American Fisheries Society 69:135–140.

Shaw, P. A., and J. A. Maga. 1943. The effect of mining silt on yield of fry from salmon spawning beds. California Fish and Game 29:29–41.

Shelton, J. M. 1955. The hatching of chinook salmon eggs under simulated stream conditions. Progressive Fish-Culturist 17:20–35.

Shelton, J. M., and R. D. Pollock. 1966. Siltation and egg survival in incubation channels. Transactions of the American Fisheries Society 95:183–187.

Shepard, B. B., S. A. Leathe, T. M. Weaver, and M. D. Enk. 1984. Monitoring levels of fine sediment within tributaries to Flathead Lake, and impacts of fine sediment on bull trout recruitment. Pages 146–156 in F. Richardson and R. H. Hamre, editors. Wild trout III. Federation of Fly Fishers and Trout Unlimited, Vienna, Virginia.

Sheridan, W. L. 1962. Waterflow through a salmon spawning riffle in southeastern Alaska. U.S. Fish and Wildlife Service Special Scientific Report Fisheries 407.

Sheridan, W. L., and W. J. McNeil. 1968. Some effects of logging on two salmon streams in Alaska. Journal of Forestry 66:128–133.

Sheridan, W. L., M. P. Perensovich, T. Faris, and K. Koski. 1984. Sediment content of streambed gravels in some pink salmon spawning streams in Alaska. Pages 153–165 *in* Meehan et al. (1984).

Shields, H. J. 1968. Riffle sifter for Alaska salmon gold. Pages 204–208 *in* Science for better understanding. U.S. Department of Agriculture Yearbook of Agriculture.

Shiozawa, D. K. 1986. The seasonal community structure and drift of microcrustaceans in Valley Creek, Minnesota. Canadian Journal of Zoology 64:1655–1664.

Shirazi, M. A., and W. K. Seim. 1979. A stream systems evaluation: an emphasis on spawning habitat for salmonids. U.S. Environmental Protection Agency, EPA Report 600/3-79-109, Washington, DC.

Sigler, J. W., T. C. Bjornn, and F. H. Everest. 1984. Effects of chronic turbidity on density and growth of steelheads and coho salmon. Transactions of the American Fisheries Society 113:142–150.

Simanton, J. R., G. D. Wingate, and M. A. Weltz. 1990. Runoff and sediment from a burned sagebrush community. Pages 180–185 *in* J. S. Krammes, editors. Effects of fire management of southwestern natural resources. U.S. Forest Service General Technical Report RM-191.

Simmons, C. E. 1976. Sediment characteristics of streams in the eastern Piedmont and western coastal plain regions of North Carolina. U S. Geological Survey, Water Supply Paper 1798–0, Washington, DC.

Simons, D. B. 1979. Effects of stream regulation on channel morphology. Pages 95–111 *in* J. V. Ward and J. A. Stanford, editors. The ecology of regulated streams. Plenum Press, New York.

Simons, D. B., and R-M. Li. 1983. Modeling of sediment nonpoint source pollution from watersheds. Pages 341–373 *in* Overcash and Davidson (1983).

Slaney, P. A., T. G. Halsey, and H. A. Smith. 1977a. Some effects of forest harvesting on salmonid rearing habitat in two streams in the central interior of British Columbia. British Columbia Fish and Wildlife Branch, Fisheries Management Report 71, Victoria.

Slaney, P. A., T. G. Halsey, and A. F. Tautz. 1977b. Effects of forest harvesting practices on spawning habitat of stream salmonids in the Centenial Creek watershed, British Columbia. British Columbia Fish and Wildlife Branch, Fisheries Management Report 73, Victoria.

Sloane-Richey, J., M. A. Perkins, and K. W. Malueg. 1981. The effects of urbanization and stormwater runoff on the food quality in two salmonid streams. Proceedings of the International Association of Theoretical and Applied Limnology 21:812–818.

Smith, O. R. 1940. Placer mining silt and its relation to salmon and trout on the Pacific coast. Transactions of the American Fisheries Society 69:225–230.

Smith, P. W. 1971. Illinois streams: a classification based on their fishes and analysis of factors responsible for disappearance of native species. Illinois Natural History Survey Biological Notes 76.

Somer, W. L., and T. J. Hassler. 1992. Effects of suction-dredge gold mining on benthic invertebrates in a northern California stream. North American Journal of Fisheries Management 12:244–252.

Sopper, W. E., and H. W. Lull, editors. 1967. Forest hydrology: proceedings of a symposium. Pergamon Press, New York.

Sorensen, D. L., M. M. McCarthy, E. J. Middlebrooks, and D. B. Porcella. 1977. Suspended and dissolved solids effects on freshwater biota: a review. U.S. Environmental Protection Agency, EPA Report 600/3-77-042, Washington, DC.

Sowden, T. K., and G. Power. 1985. Prediction of rainbow trout embryo survival in relation to groundwater seepage and particle size of spawning substrates. Transactions of the American Fisheries Society 114:804–812.

Spaulding, W. M., Jr., and R. D. Ogden. 1968. Effects of surface mining on the fish and wildlife resources of the United States. U.S. Fish and Wildlife Service Bureau of Sport Fisheries and Wildlife Publication 68.

Sprules, W. M. 1947. An ecological investigation of stream insects in Algonquin Park, Ontario. University of Toronto Studies, Biological Series 56. Publication of the Ontario Fisheries Research Laboratory 69:1–80.

Starnes, L. B. 1983. Effects of surface mining on aquatic resources in North America. Fisheries 8(6):2–4.

Starnes, L. B. 1985. Aquatic community response to techniques utilized to reclaim eastern U.S. coal surface mine-impacted streams. Pages 193–222 *in* Gore (1985a).

Stein, C. B. 1972. Population changes in the naiad mollusk fauna of the lower Olentangy River following channelization and highway construction. Bulletin of the American Malacological Union Incorporated February:47–49.

Steinblums, I. J., H. A. Froehlich, and J. K. Lyons. 1984. Designing stable buffer strips for stream protection. Journal of Forestry 82:49–52.

Stichling, W. 1974. Sediment loads in Canadian rivers. Environment Canada, Water Resources Branch, Technical Bulletin 74, Ottawa.

Stowell, R., A. Espinosa, T. C. Bjornn, W. S. Platts, D. C. Burns, and J. S. Irving. 1983. Guide for predicting salmonid response to sediment yields in Idaho batholith watersheds. U.S. Forest Service, Intermountain Region, Ogden, Utah.

Striffler, W. D. 1964. Sediment, streamflow, and land use relationships in northern lower Michigan. U.S. Forest Service, Lake States Forest Experiment Station, St. Paul, Minnesota, and Michigan Department of Conservation, Research Paper LS-16, Ann Arbor, Michigan.

Stuart, T. A. 1953. Spawning migration, reproduction and young stages of loch trout (*Salmo trutta* L.). Scottish Home Department, Freshwater and Salmon Fisheries Research 5, Edinburgh, UK.

Stuber, R. J. 1985. Trout habitat, abundance, and fishing opportunities in fenced vs unfenced riparian habitat along Sheep Creek, Colorado. Pages 310–314 *in* Johnson et al. (1985).

Sullivan, K., T. E. Lisle, C. A. Dolloff, G. E. Grant, and L. M. Reid. 1987. Stream channels: the link between forests and fishes. Pages 39–97 *in* Salo and Cundy (1987).

Sumner, F. H., and O. R. Smith. 1940. Hydraulic mining and debris dams in relation to fish life in the American and Yuba rivers of California. California Fish and Game 26:2–22.

Swank, W. T., and D. A. Crossley, Jr., editors. 1988. Forest hydrology and ecology at Coweeta. Springer-Verlag, New York.

Swanston, D. N. 1991. Natural processes. Pages 139–179 *in* Meehan (1991a).

Swift, L. W., Jr. 1984a. Gravel and grass surfacing reduces soil loss from mountain roads. Forest Science 30:657–670.

Swift, L. W., Jr. 1984b. Soil losses from roadbeds and cut and fill slopes in the southern Appalachian Mountains. Southern Journal of Applied Forestry 8:209–216.

Swift, L. W., Jr. 1985. Forest road design to minimize erosion in the southern Appalachians. Pages 141–151 *in* Blackmon (1985).

Swift, L. W., Jr. 1986. Filter strip widths for forest roads in the southern Appalachians. Southern Journal of Applied Forestry 10:27–34.

Swift, L. W., Jr. 1988. Forest access roads, design, maintenance, and soil loss. Pages 313–324 *in* Swank and Crossley (1988).

Sykora, J. L, E. J. Smith, and M. Synak. 1972. Effect of lime neutralized iron hydroxide suspensions on juvenile brook trout (*Salvelinus fontinalis* Mitchill). Water Research 6:935–950.

Taft, A. C., and L. Shapovalov. 1935. A biological survey of streams and lakes in the Klamath and Shasta national forests of California. U.S. Bureau of Fisheries, Washington, DC. (Not seen; cited in Cordone and Kelley 1961).

Tagart, J. V. 1984. Coho salmon survival from egg deposition to fry emergence. Pages 173–182 *in* J. M. Walton and D. B. Houston, editors. Proceedings of the Olympic Wild Fish Conference. Peninsula College, Port Angeles, Washington.

Tappel, P. D., and T. C. Bjornn. 1983. A new method of relating size of spawning gravel to salmonid embryo survival. North American Journal of Fisheries Management 3:123–135.

Tarzwell, C. M. 1938. Factors influencing fish food and fish production in southwestern streams. Transactions of the American Fisheries Society 67:246–255.

Tebo, L. B., Jr. 1955. Effects of siltation, resulting from improper logging, on the bottom fauna of a small trout stream in the southern Appalachians. Progressive Fish-Culturist 17:64–70.

Tebo, L. B., Jr. 1957. Effects of siltation on trout streams. Proceedings of the Society of American Foresters 1956:198–202.

Tennant, D. L. 1976. Instream flow regimens for fish, wildlife, recreation and related environmental resources. Fisheries 1(4):6–10.

Thames, J. L., editor. 1977. Reclamation and use of disturbed land in the Southwest. University of Arizona Press, Tucson.

Thirgood, J. V. 1978. Extent of disturbed land and major reclamation problems in Canada. Pages 45–68 in Schaller and Sutton (1978).

Thomas, V. G. 1985. Experimentally determined impacts of a small, suction gold dredge on a Montana stream. North American Journal of Fisheries Management 5:480–488.

Thorn, W. C. 1988. Evaluation of habitat improvement for brown trout in agriculturally damaged streams of southeastern Minnesota. Minnesota Department of Natural Resources Section of Fisheries Investigational Report 394.

Thornburg, A. A., and S. H. Fuchs. 1978. Plant materials and requirements for growth in dry regions. Pages 411–423 in Schaller and Sutton (1978).

Thorne, C. R. 1982. Processes and mechanisms of river bank erosion. Pages 227–271 in Hey et al. (1982).

Thorne, C. R., J. C. Bathurst, and R. D. Hey, editors. 1987. Sediment transport in gravel-bed rivers. Wiley, New York.

Titcomb, J. W. 1926. Forests in relation to freshwater fishes. Transactions of the American Fisheries Society 56:122–129.

Trautman, M. B. 1933. The general effects of pollution on Ohio fish life. Transactions of the American Fisheries Society 63:69–72.

Trautman, M. B. 1981. The fishes of Ohio, 2nd edition. Ohio State University Press, Columbus.

Trimble, G. R., Jr., and R. S. Sartz. 1957. How far from a stream should a logging road be located? Journal of Forestry 55:339–341.

Trimble, S. W., and J. C. Knox. 1984. Comment on 'Erosion, redeposition, and delivery of sediment to midwestern streams' by D. C. Wilkin and S. J. Hebel. Water Resources Research 20:1317–1318.

Trimble, S. W., and S. W. Lund. 1982. Soil conservation and the reduction of erosion and sedimentation in the Coon Creek basin, Wisconsin. U.S. Geological Survey, Professional Paper 1234, Washington, DC.

Tripp, D. B., and V. A. Poulin. 1986. The effects of mass wasting on juvenile fish habitats in streams on the Queen Charlotte Islands. British Columbia Ministry of Forests and Lands, Land Management Report 45, Victoria.

Troelstrup, N. H., Jr., and J. A. Perry. 1989. Water quality in southeastern Minnesota streams: observations along a gradient of land use and geology. Journal of the Minnesota Academy of Science 55:6–13.

Truhlar, J. F. 1976. Determining suspended sediment loads from turbidity records. Pages 7/65-7/74 in USWRC (1976).

Tryon, C. P. 1976. Excavated sediment traps prove superior to dammed ones. Pages 2/42-2/51 in USWRC (1976).

Tsui, P. T. P., and P. J. McCart. 1981. Effects of stream-crossing by a pipeline on the benthic macroinvertebrate communities of a small mountain stream. Hydrobiologia 79:271–276.

Ursic, S. J. 1991. Hydrologic effects of clearcutting and stripcutting loblolly pine in the Coastal Plain. Water Resources Bulletin 27:925–937.

U.S. Army Corps of Engineers, and eleven agencies. 1961. The single-stage sampler for suspended sediment. Inter-Agency Committee on Water Resources Report 13, U.S. Government Printing Office, Washington, DC.

U.S. Bureau of Reclamation. 1948. Proceedings of the federal inter-agency sedimentation conference, 1946. U.S. Department of the Interior, Bureau of Reclamation, Denver, Colorado. (Not seen; cited in USWRC 1976).

USDA (U.S. Department of Agriculture). 1955. Water. U.S. Department of Agriculture Yearbook of Agriculture.

USDA (U.S. Department of Agriculture). 1965. Proceedings of the federal inter-agency sedimentation conference, 1963. U.S. Department of Agriculture Miscellaneous Publication 970.

USDA (U.S. Department of Agriculture). 1969. Control of agriculture related pollution: a report to the President. U.S. Department of Agriculture, Office of Science and Technology, Washington, DC. (Not seen; cited in Oschwald 1972.)

USDI (U.S. Department of the Interior). 1990. Status of the Trinity River restoration program. U.S. Bureau of Reclamation and U.S. Fish and Wildlife Service, Weaverville, California.

USEPA (U.S. Environmental Protection Agency). 1973a. Methods and practices for controlling water pollution from agricultural nonpoint sources. U.S. Environmental Protection Agency, EPA Report 430/9-73-015, Washington, DC.

USEPA (U.S. Environmental Protection Agency). 1973b. Processes, procedures, and methods to control pollution resulting from all construction activity. U.S. Environmental Protection Agency, EPA Report 430/9-73-007, Washington, DC.

USEPA (U.S. Environmental Protection Agency). 1990. The quality of our nation's water: a summary of the 1988 National Water Quality Inventory. U.S. Environmental Protection Agency, EPA Report 440/4-90-005, Washington, DC.

USEPA (U.S. Environmental Protection Agency). 1993. Guidance specifying management measures for sources of nonpoint pollution in coastal waters. U.S. Environmental Protection Agency, EPA Report 840-B-92-002, Washington, DC.

USFS (U.S. Forest Service). 1975. Symposium on trout habitat research and management proceedings. U.S. Forest Service, Southeastern Forest Experiment Station, Asheville, North Carolina.

USFS (U.S. Forest Service). 1977. Anatomy of a mine from prospect to production. U.S. Forest Service General Technical Report INT-35.

USFS (U.S. Forest Service). 1979a. User guide to vegetation: mining and reclamation in the West. U.S. Forest Service General Technical Report INT-64.

USFS (U.S. Forest Service). 1979b. User guide to soils: mining and reclamation in the West. U.S. Forest Service General Technical Report INT-68.

USFS (U S. Forest Service). 1980. User guide to hydrology: mining and reclamation in the West. U.S. Forest Service General Technical Report INT-74.

USFWS (U.S. Fish and Wildlife Service). 1977. Proceedings of a seminar: improving fish and wildlife benefits in range management. U.S. Fish and Wildlife Service FWS-OBS-77/1.

USSCS (U.S. Soil Conservation Service). 1973. Guides for erosion & sediment control in developing areas of New Hampshire. New Hampshire Association of Conservation Districts, and the North Country Resource Conservation and Development Project, Durham, New Hampshire.

USSCS (U.S. Soil Conservation Service). 1982. Ponds—planning, design, construction. U.S. Soil Conservation Service, Washington, DC.

USSCS (U.S. Soil Conservation Service). 1983. National engineering handbook, section 3. Sedimentation. U.S. Department of Agriculture, Washington, DC.

USWRC (U.S. Water Resources Council). 1976. Proceedings of the third federal inter-agency sedimentation conference, 1976. U.S. Water Resources Council, Denver, Colorado.

Van Nieuwenhuyse, E. E., and J. D. LaPerriere. 1986. Effects of placer gold mining on primary production in subarctic streams of Alaska. Water Resources Bulletin 22:91–99.

Van Velson, R. 1979. Effects of livestock grazing upon rainbow trout in Otter Creek, Nebraska. Pages 53–55 *in* Cope (1979).

Vannote, R. L., G. W. Minshall, K. W. Cummins, J. R. Sedell, and C. E. Cushing. 1980. The river continuum concept. Canadian Journal of Fisheries and Aquatic Sciences 37:130–137.

Verma, T. R., and J. L. Thames. 1978. Grading and shaping for erosion control and vegetative establishment in dry regions. Pages 399–409 *in* Schaller and Sutton (1978).

Vice, R. B., H. P. Guy, and C. E. Ferguson. 1969. Sediment movement in an area of suburban highway construction, Scott Run Basin, Fairfax County, Virginia, 1961–64. U.S. Geological Survey Water Supply Paper 1591-E, Washington, DC.

Vohs, P. A., I. J. Moore, and J. S. Ramsey. 1993. A critical review of the effects of turbidity on aquatic organisms in large rivers. U.S. Fish and Wildlife Service, Environmental Management Technical Center, Report 93-S002, Onalaska, Wisconsin.

Vories, K. D., editor. 1976. Reclamation of western surface mined lands. Ecology Consultants, Inc., Fort Collins, Colorado.

Vowell, J. L. 1985. Erosion rates and water quality impacts from a recently established forest road in Oklahoma's Ouachita Mountains. Pages 152–163 *in* Blackmon (1985).

Waddell, T. E., editor. 1986. The off-site costs of soil erosion: proceedings of a symposium. The Conservation Foundation, Washington, DC.

Wagener, S. M., and J. D. LaPerriere. 1985. Effects of placer mining on the invertebrate communities of interior Alaska streams. Freshwater Invertebrate Biology 4:208–214.

Walkotten, W. J. 1973. A freezing technique for sampling streambed gravel. U.S. Forest Service Research Note PNW-205.

Walkotten, W. J. 1976. An improved technique for freeze sampling streambed sediments. U.S. Forest Service Research Note PNW-281.

Wallace, J. B., and M. E. Gurtz. 1986. Response of *Baetis* mayflies (Ephemeroptera) to catchment logging. American Midland Naturalist 115:25–41.

Wallace, J. B., and R. W. Merritt. 1980. Filter-feeding ecology of aquatic insects. Annual Review of Entomology 25:103–132.

Wallace, R., and editors of Time-Life Books. 1976. The miners. Time-Life Books, Alexandria, Virginia.

Wallen, I. E. 1951. The direct effect of turbidity on fishes. Bulletin of the Oklahoma Agricultural Experiment Station 48(2):1–27.

Walling, D. E., and P. Kane. 1984. Suspended sediment properties and their geomorphological significance. Pages 311–334 *in* T. P. Burt and D. E. Walling, editors. Catchment experiments in fluvial geomorphology. GEOBooks, Norwich, UK.

Ward, H. B. 1937/1938. Placer mining on the Rogue River, Oregon, in its relation to the fish and fishing in that stream. Oregon Department of Geology and Mineral Industries, Portland.

Ward, T. J., and A. D. Seiger. 1983. Adaptation and application of a surface erosion model for New Mexico forest roadways. New Mexico Water Resources Research Institute, Santa Fe.

Wark, J. W., and F. J. Keller. 1963. Preliminary study of sediment sources and transport in Potomac River basin. U.S. Geological Survey and Interstate Commission on the Potomac River Basin, Washington, DC.

Waters, T. F. 1965. Interpretation of invertebrate drift in streams. Ecology 46:327–334.

Waters, T. F. 1972. The drift of stream insects. Annual Review of Entomology 17:253–272.

Waters, T. F. 1982. Annual production by a stream brook charr population and by its principal invertebrate food. Environmental Biology of Fishes 7:165–170.

Waters, T. F. 1984. Annual production by *Gammarus pseudolimnaeus* among substrate types in Valley Creek, Minnesota. American Midland Naturalist 112:95–102.

Waters, T. F. 1988. Fish production-benthos production relationships in trout streams. Polskie Archiwum Hydrobiologii 35:545–561.

Waters, T. F. 1993. Dynamics in stream ecology. Pages 1–8 *in* R. J. Gibson and R. E. Cutting, editors. Production of juvenile Atlantic salmon, *Salmo salar*, in natural waters. Canadian Special Publication of Fisheries and Aquatic Sciences 118.

Waters, T. F., and J. C. Hokenstrom. 1980. Annual production and drift of the stream amphipod *Gammarus pseudolimnaeus* in Valley Creek, Minnesota. Limnology and Oceanography 25:700–710.

WDNR (Wisconsin Department of Natural Resources). 1992. Wisconsin construction site best management practice handbook (revised). Wisconsin Department of Natural Resources, Bureau of Water Resources Management, Madison.

Weber, P. K. 1986. The use of chemical flocculants for water clarification: a review of the literature with application to placer mining. Alaska Department of Fish and Game, Technical Report 86–4, Juneau.

Webster, D. A., and G. Eiriksdottir. 1976. Upwelling water as a factor influencing choice of spawning sites by brook trout (*Salvelinus fontinalis*). Transactions of the American Fisheries Society 105:416–421.

Webster, J. R., and six coauthors. 1983. Stability of stream ecosystems. Pages 355–395 *in* Barnes and Minshall (1983).

Welch, H. E., P. E. K. Symons, and D. W. Narver. 1977. Some effects of potato farming and forest clearcutting on small New Brunswick streams. New Brunswick Fisheries and Marine Service, Technical Report 745, St. Andrews.

Wentworth, C. K. 1922. A scale of grade and class terms for clastic sediments. Journal of Geology 30:377–392.

Wesche, T. A. 1985. Stream channel modifications and reclamation structures to enhance fish habitat. Pages 103–163 *in* Gore (1985a).

Wesche, T. A., V. R. Hasfurther, W. A. Hubert, and Q. D. Skinner. 1987. Assessment of flushing flow recommendations in a steep, rough, regulated tributary. Pages 59–69 *in* Craig and Kemper (1987).

West, J. L. 1979. Intragravel characteristics in some western North Carolina trout streams. Proceedings of the Annual Conference Southeastern Association of Fish and Wildlife Agencies 32(1978):625–633.

West, J. L., G. S. Grindstaff, C. McIlwain, P. G. White, and R. Bacon. 1982. A comparison of trout populations, reproductive success, and characteristics of a heavily silted and a relatively unsilted stream in western North Carolina. Final report (Contract C-1054) to North Carolina Department of Natural Resources and Community Development, Raleigh.

White, D. S., and J. R. Gammon. 1977. The effect of suspended solids on macroinvertebrate drift in an Indiana creek. Proceedings of the Indiana Academy of Science 86:182–188.

White, R. J. 1975. In-stream management for wild trout. Pages 48–58 *in* W. King, editor. Proceedings of the wild trout management symposium. Trout Unlimited, Vienna, Virginia.

White, R. J., and O. M. Brynildson. 1967. Guidelines for management of trout stream habitat in Wisconsin. Wisconsin Department of Natural Resources Technical Bulletin 39.

Whiting, E. R., and H. F. Clifford. 1983. Invertebrates and urban runoff in a small northern stream, Edmonton, Alberta, Canada. Hydrobiologia 102:73–80.

Whitman, R. P., T. P. Quinn, and E. L. Brannon. 1982. Influence of suspended volcanic ash on homing behavior of adult chinook salmon. Transactions of the American Fisheries Society 111:63–69.

Whitworth, M. R., and D. C. Martin. 1990. Instream benefits of CRP filter strips. Transactions of the North American Wildlife and Natural Resources Conference 55:40–45.

Wickett, W. P. 1954. The oxygen supply to salmon eggs in spawning beds. Journal of the Fisheries Research Board of Canada 11:933–953.

Wickett, W. P. 1958. Review of certain environmental factors affecting the production of pink and chum salmon. Journal of the Fisheries Research Board of Canada 15:1103–1126.

Wilkin, D. C., and S. J. Hebel. 1982. Erosion, redeposition, and delivery of sediment to midwestern streams. Water Resources Research 18:1278–1282.

Wilson, J. N. 1957. Effects of turbidity and silt on aquatic life. Pages 235–239 *in* C. M. Tarzwell, editor. Transactions of a seminar on biological problems in water pollution. U.S. Public Health Service, Robert A. Taft Sanitation Engineering Center, Cincinnati, Ohio.

Wilson, J. N. 1959. The effects of erosion, silt, and other inert materials on aquatic life. Pages 269–271 *in* C. M. Tarzwell, editor. Transactions of the second seminar on biological problems in water pollution. U. S. Public Health Service, Robert A. Taft Sanitary Engineering Center, Cincinnati, Ohio.

Wilson, M. P. 1983. Erosion of banks along Piedmont urban streams. University of North Carolina, Water Resources Research Institute Report 189, Charlotte.

Winegar, H. H. 1977. Camp Creek channel fencing—plant, wildlife, soil, and water response. Rangeman's Journal 4:10–12.

Wittmuss, H. D. 1987. Erosion control incorporating conservation tillage, crop rotation and structural practices. Pages 90–99 *in* Optimum erosion control at least cost. American Society of Agricultural Engineers, St. Joseph, Michigan.

Witzel, L. D., and H. R. MacCrimmon. 1981. Role of gravel substrate and ova survival and alevin emergence of rainbow trout, *Salmo gairdneri*. Canadian Journal of Zoology 59:629–636.

Witzel, L. D., and H. R. MacCrimmon. 1983a. Embryo survival and alevin emergence of brook char, *Salvelinus fontinalis*, and brown trout, *Salmo trutta*, relative to redd gravel composition. Canadian Journal of Zoology 61:1783–1792.

Witzel, L. D., and H. R. MacCrimmon. 1983b. Redd-site selection by brook trout and brown trout in southwestern Ontario streams. Transactions of the American Fisheries Society 112:760–771.

Wolf, Ph. 1950. American problems and practice. I. Salmon which disappeared. Salmon and Trout Magazine 130:201–212.

Wolman, M. G. 1964. Problems posed by sediment derived from construction activities in Maryland. Report to the Maryland Water Pollution Control Commission, Annapolis, Maryland.

Wolman, M. G., and A. P. Schick. 1967. Effects of construction on fluvial sediment, urban and suburban areas of Maryland. Water Resources Research 3:451–464.

Woodward-Clyde Consultants. 1980a. Gravel removal studies in Arctic and Subarctic floodplains in Alaska. U.S. Fish and Wildlife Service FWS-OBS-80/08.

Woodward-Clyde Consultants. 1980b. Gravel removal guidelines manual for Arctic and Subarctic floodplains. U.S. Fish and Wildlife Service FWS-OBS-80/09.

Wright, D. L., H. D. Perry, and R. E. Blaser. 1978. Persistent low maintenance vegetation for erosion control and aesthetics in highway corridors. Pages 553–583 *in* Schaller and Sutton (1978).

WWRC (Washington Water Research Center). 1981. Proceedings from the conference on salmon-spawning gravel: a renewable resource in the Pacific Northwest. Washington State University, Washington Water Research Center, Report 39, Pullman.

Yamamoto, T. 1982. A review of uranium spoil and mill tailings revegetation in the western United States. U.S. Forest Service General Technical Report RM-92.

Yee, C. S., and T. D. Roelofs. 1980. Planning forest roads to protect salmonid habitat, part 4. Influence of forest and rangeland management on anadromous fish habitat in western North America. U.S. Forest Service General Technical Report PNW-109.

Yorke, T. H., and W. J. Herb. 1978. Effects of urbanization on streamflow and sediment transport in the Rock Creek and Anacostia River basins, Montgomery County, Maryland, 1962–74. U.S. Geological Survey, Professional Paper 1003, Washington, DC.

Young, M. K., W. A. Hubert, and T. A. Wesche. 1989. Substrate alteration by spawning brook trout in a southeastern Wyoming stream. Transactions of the American Fisheries Society 118:379–385.

Young, M. K., W. A. Hubert, and T. A. Wesche. 1991. Selection of measures of substrate composition to estimate survival to emergence of salmonids and to detect changes in stream substrates. North American Journal of Fisheries Management 11:339–346.

Young, R. A., and C. A. Onstad. 1987. Soil erosion: measurement and prediction. Pages 91–111 *in* Harlin and Berardi (1987).

Yount, J. D., and G. J. Niemi. 1990. Recovery of lotic communities and ecosystems following disturbance: theory and application. Environmental Management 14:515–762.

APPENDIX I

SPECIAL BIBLIOGRAPHY ON REVIEWS, BIBLIOGRAPHIES, AND REFERENCE BOOKS

Anderson, N., and J. Sedell. 1979. Detritus processing by macroinvertebrates in stream ecosystems. Annual Review of Entomology 24:351–377.

Armour, C. L., D. A. Duff, and W. Elmore. 1991. The effects of livestock grazing on riparian and stream ecosystems. Fisheries 16(1):7–11.

Baker, R. A., editor. 1980a. Contaminants and sediments, volume 1. Fate and transport, case studies, modeling, toxicity. Ann Arbor Science, Ann Arbor, Michigan.

Baker, R. A., editor. 1980b. Contaminants and sediments, volume 2. Analysis, chemistry, biology. Ann Arbor Science, Ann Arbor, Michigan.

Barnes, J. R., and G. W. Minshall, editors. 1983. Stream ecology: application and testing of general ecological theory. Plenum Press, New York.

Batie, S. S. 1983. Soil erosion: crisis in America's croplands? The Conservation Foundation, Washington, DC.

Baudo, R., J. P. Giesy, and H. Muntau. 1990. Sediments: chemistry and toxicity of in-place pollutants. Lewis Publishers, Ann Arbor, Michigan.

Beasley, R. P., J. M. Gregory, and T. R. McCarty. 1984. Erosion and sediment pollution control, 2nd edition. Iowa State University Press, Ames.

Bjornn, T. C., and D. W. Reiser. 1991. Habitat requirements of salmonids in streams. Pages 83–138 in Meehan (1991a).

Bjornn, T.C., and seven coauthors. 1974. Sediment in streams and its effects on aquatic life. University of Idaho, Water Resources Research Institute, Research Technical Completion Report, Project B-025-IDA, Moscow.

Bjornn, T. C., and six coauthors. 1977. Transport of granitic sediment in streams and its effects on insects and fish. University of Idaho, College of Forestry, Wildlife and Range Sciences, Bulletin 17, Moscow.

Black, P. E. 1991. Watershed hydrology. Prentice Hall, Englewood Cliffs, New Jersey.

Blauch, B. W. 1978. Reclamation of lands disturbed by stone quarries, sand and gravel pits, and borrow pits. Pages 619–628 in Schaller and Sutton (1978).

Campbell, I. C., and T. J. Doag. 1989. Impact of timber harvesting and production on streams: a review. Australian Journal of Marine and Freshwater Research 40:519–539.

Chapman, D. W. 1988. Critical review of variables used to define effects of fines in redds of large salmonids. Transactions of the American Fisheries Society 117:1–21.

Cordone, A. J., and D. W. Kelley. 1961. The influences of inorganic sediment on the aquatic life of streams. California Fish and Game 47:189–228.

Cummins, K. W. 1962. An evaluation of some techniques for the collection and analysis of benthic samples with special emphasis on lotic waters. American Midland Naturalist 67:477–504.

Cummins, K. W. 1966. A review of stream ecology with special emphasis on organism-substrate relationships. Pages 2–51 *in* Cummins et al. (1966).

Cummins, K. W. 1974. Structure and function of stream ecosystems. BioScience 24:631–641.

Cummins, K. W., and M. J. Klug. 1979. Feeding ecology of stream invertebrates. Annual Review of Ecology and Systematics 10:147–172.

Dieter, C. D. 1990. Causes and effects of water turbidity: a selected annotated bibliography. South Dakota State University, South Dakota Cooperative Wildlife Research Unit, Technical Bulletin 5, Brookings.

Duff, D. A., N. Banks, E. Sparks, W. E. Stone, and R. J. Poehlmann. 1988. Indexed bibliography on stream habitat improvement (4th revision). U.S. Forest Service, Intermountain Region, Ogden, Utah.

EIFAC (European Inland Fisheries Advisory Commission). 1965. Water quality criteria for European freshwater fish. Report on finely divided solids and inland fisheries. International Journal of Air and Water Pollution 9:151–168.

Elwood, J. W., J. D. Newbold, R. V. O'Neill, and W. Van Winkle. 1983. Resource spiraling: an operational paradigm for analyzing lotic ecosystems. Pages 3–28 *in* T. D. Fontaine, III, and S. M. Bartell, editors. Dynamics of lotic ecosystems. Ann Arbor Science, Ann Arbor, Michigan.

Everest, F. H., R. L. Beschta, J. C. Scrivener, K. V. Koski, J. R. Sedell, and C. J. Cederholm. 1987. Fine sediment and salmonid production: a paradox. Pages 98–142 *in* Salo and Cundy (1987).

Farnworth, E. G., and seven coauthors. 1979. Impacts of sediment and nutrients on biota in surface waters of the United States. U.S. Environmental Protection Agency, EPA Report 600/3-79-105, Washington, DC.

Gammon, J. R. 1970. The effect of inorganic sediment on stream biota. U.S. Environmental Protection Agency, Water Quality Office, Grant Number 18050 DWC, Washington, DC.

Garman, G. C., and J. R. Moring. 1987. Impacts of sedimentation on aquatic systems: a literature review and proposed research agenda. Report prepared for the Maine Land Use Regulation Commission, Committee on Stream Research, Orono.

Gibbons, D. R., W. R. Meehan, K. V. Koski, and T. R. Merrell, Jr. 1987. History of studies of fisheries and forestry interactions in southeastern Alaska. Pages 297–329 *in* Salo and Cundy (1987).

Gibbons, D. R., and E. O. Salo. 1973. An annotated bibliography of the effects of logging on fish of the western United States and Canada. U.S. Forest Service, General Technical Report PNW-10.

Gordon, N. D., T. A. McMahon, and B. L. Finlayson. 1992. Stream hydrology: an introduction for ecologists. Wiley, New York.

Gore, J. A., editor. 1985a. The restoration of rivers and streams: theories and experience. Butterworth, Boston.

Gregory, J. D., and J. L. Stokoe. 1981. Streambank management. Pages 276–281 *in* Krumholz (1981).

Hall, T. J. 1984a. The effects of fine sediment on salmonid spawning gravel and juvenile rearing habitat: a literature review. National Council of the Paper Industry for Air and Stream Improvement, Technical Bulletin 428, New York.

Hall, T. J., R. K. Haley, J. C. Gislason, and G. G. Ice. 1984. The relationship between fine sediment and macroinvertebrate community characteristics: a literature review and results from NCASI fine sediment studies. National Council of the Paper Industry for Air and Stream Improvement, Technical Bulletin 418, New York.

Hall, T. J., and G. G. Ice. 1981. Forest management practices, sediment load characterization, streambed composition and predicting aquatic organism response: where we are and where we need to go. National Council of the Paper Industry for Air and Stream Improvement, Technical Bulletin 353, New York.

Harlin, J. M., and G. M. Berardi, editors. 1987. Agricultural soil loss: processes, policies, and prospects. Westview Press, Boulder, Colorado.

Harmon, W. E., and twelve coauthors. 1986. Ecology of coarse woody debris in temperate ecosystems. Advances in Ecological Research 15:133–302.

Hassett, J. J., and W. L. Banwart. 1992. Soils and their environment. Prentice Hall, Englewood Cliffs, New Jersey.

Hey, R. D., J. C. Bathurst, and C. R. Thorne, editors. 1982. Gravel-bed rivers: fluvial processes, engineering and management. Wiley, New York.

Hynes, H. B. N. 1966. The biology of polluted waters. Liverpool University Press, Liverpool, UK.

Hynes, H. B. N. 1970. The ecology of running waters. University of Toronto Press, Toronto.

Hynes, H. B. N. 1973. The effects of sediment on the biota in running water. Pages 653–663 in Fluvial processes and sedimentation. Canadian Department of the Environment, Ottawa.

Iwamoto, R. N., E. O. Salo, M. A. Madej, and R. L. McComas. 1978. Sediment and water quality: a review of the literature including a suggested approach for water quality criteria. U.S. Environmental Protection Agency, EPA Report 910/9-78-048, Washington, DC.

Karr, J. R., L. A. Toth, and D. R. Dudley. 1985. Fish communities of midwestern rivers: a history of degradation. BioScience 35(2):90–95.

Kauffman, J. B., and W. C. Krueger. 1984. Livestock impacts on riparian ecosystems and streamside management implications . . . a review. Journal of Range Management 37:430–438.

Lenat, D. R. 1984. Agriculture and stream water quality: a biological evaluation of erosion control practices. Environmental Management 8:333–344.

Lenat, D. R., D. L. Penrose, and K. W. Eagleson. 1979. Biological evaluation of non-point source pollutants in North Carolina streams and rivers. North Carolina Department of Natural Resources and Community Development, Biological Series 102, Raleigh.

Leopold, L. B. 1994. A view of the river. Harvard University Press, Cambridge, Massachusetts.

Leopold, L. B., M. G. Wolman, and J. P. Miller. 1964. Fluvial processes in geomorphology. Freeman, San Francisco.

Lloyd, D. S. 1985. Turbidity in freshwater habitats of Alaska: a review of published and unpublished literature relevant to the use of turbidity as a water quality standard. Alaska Department of Fish and Game, Habitat Division, Report 85-1, Juneau.

Lloyd, D. S. 1987. Turbidity as a water quality standard for salmonid habitats in Alaska. North American Journal of Fisheries Management 7:34–45.

Macdonald, J. S., G. Miller, and R. A. Stewart. 1988. The effects of logging, other forest industries and forest management practices on fish: an initial bibliography. Canadian Technical Report of Fisheries and Aquatic Sciences 1622.

Meehan, W. R., editor. 1991a. Influences of forest and rangeland management on salmonid fishes and their habitats. American Fisheries Society, Special Publication 19.

Mih, W. C. 1978. A review of restoration of stream gravel for spawning and rearing of salmon species. Fisheries 3(1):16–18.

Miller, E. L., R. S. Beasley, and J. C. Covert. 1985. Forest road sediments: production and delivery to streams. Pages 164–176 in Blackmon (1985).

Minshall, G. W. 1988. Stream ecosystem theory: a global perspective. Journal of the North American Benthological Society 7:263–288.

Moring, J. R., and G. C. Garman. 1986. The value of riparian zones for fisheries. Pages 81–90 in J. A. Bissonette, editor. Is good forestry good wildlife management? University of Maine at Orono, Maine Agricultural Experiment Station Miscellaneous Publication 689.

Müller, K. 1974. Stream drift as a chronobiological phenomenon in running water ecosystems. Annual Review of Ecology and Systematics 5:309–323.

Muncy, R. J., G. J. Atchison, R. V. Bulkley, B. W. Menzel, L. G. Perry, and R. C. Summerfelt. 1979. Effects of suspended solids and sediment on reproduction and early life of warmwater fishes: a review. U.S. Environmental Protection Agency, EPA Report 600/3-79-042, Washington, DC.

Naiman, R. J., editor. 1992. Watershed management, balancing sustainability and environmental change. Springer-Verlag, New York.

Nelson, K. L. 1988. Review of warmwater streams habitat improvement methods. North Carolina Wildlife Resources Commission, Federal Aid in Sport Fish Restoration, Project F-22-13, Raleigh.

Newcombe, C. P., and D. D. MacDonald. 1991. Effects of suspended sediments on aquatic ecosystems. North American Journal of Fisheries Management 11:72–82.

Newport, B. D., and J. E. Moyer. 1974. State-of-the-art: sand and gravel industry. U.S. Environmental Protection Agency, EPA Report 660/2-74-066, Washington, DC.

Novinger, G. D., and J. G. Dillard, editors. 1978. New approaches to the management of small impoundments. American Fisheries Society, North Central Division, Special Publication 5, Bethesda, Maryland.

Olsen, L. A., and K. Adams. 1984. Effects of contaminated sediment on fish and wildlife: review and annotated bibliography. U.S. Fish and Wildlife Service FWS-OBS-82/66.

Overcash, M. R., and J. M. Davidson, editors. 1983. Environmental impact of nonpoint source pollution. Ann Arbor Science Publishers, Ann Arbor, Michigan.

Petticord, R. K. 1980. Direct effects of suspended sediments on aquatic organisms. Pages 501–536 *in* Baker (1980a).

Phillips, R. W. 1971. Effects of sediment on the gravel environment and fish production. Pages 64–74 *in* Krygier and Hall (1971).

Platts, W. S. 1979. Livestock grazing and riparian/stream ecosystems, an overview. Pages 39–45 *in* Cope (1979).

Platts, W. S. 1981b. Influence of forest and rangeland management on anadromous fish habitat in western North America: 7. Effects of livestock grazing. U.S. Forest Service General Technical Report PNW-124.

Platts, W. S. 1982. Livestock and riparian fishery interactions: what are the facts? Transactions of the North American Wildlife and Natural Resources Conference 47:507–515.

Platts, W. S. 1991. Livestock grazing. Pages 389–423 *in* Meehan (1991a).

Platts, W. S., and thirteen coauthors. 1983. Livestock interactions with fish and their environments. Pages 36–41 *in* Menke (1983).

Reiser, D. W., and T. C. Bjornn. 1979. Influence of forest and rangeland management on anadromous fish habitat in the western United States and Canada: 1. Habitat requirements of anadromous salmonids. U.S. Forest Service General Technical Report PNW-96.

Reiser, D. W., M. P. Ramey, and T. R. Lambert. 1985. Review of flushing flow requirements in regulated streams. Completion report to Pacific Gas and Electric Company, Department of Engineering Research, San Ramon, California.

Resh, V. H., and D. M. Rosenberg, editors. 1984. The ecology of aquatic insects. Praeger Publishers, New York.

Richardson, B. Z., and M. M. Pratt, editors. 1980. Environmental effects of surface mining of minerals other than coal: annotated bibliography and summary report. U.S. Forest Service General Technical Report INT-95.

Ryan, P. A. 1991. Environmental effects of sediment on New Zealand streams: a review. New Zealand Journal of Marine and Freshwater Research 25:207–221.

Salo, E. O., and T. W. Cundy, editors. 1987. Streamside management: forestry and fishery interactions. University of Washington Institute of Forest Resources Contribution 57.

Schwab, G. O., D. D. Fangmeier, W. J. Elliot, and R. K. Frevert. 1993. Soil and water conservation engineering, 4th edition. Wiley, New York.

Seehorn, M. E. 1987. The influence of silvicultural practices on fisheries management: effects and mitigation measures. Pages 54–63 *in* Managing southern forests for wildlife and fish, a proceedings. U.S. Forest Service General Technical Report SO-65.

Seyedbagheri, K. S., M. L. McHenry, and W. S. Platts. 1987. An annotated bibliography of the hydrology and fishery studies of the South Fork Salmon River, Idaho. U.S. Forest Service General Technical Report INT-235.

Shirazi, M. A., and W. K. Seim. 1979. A stream systems evaluation: an emphasis on spawning habitat for salmonids. U.S. Environmental Protection Agency, EPA Report 600/3-79-109, Washington, DC.

Simons, D. B. 1979. Effects of stream regulation on channel morphology. Pages 95–111 in J. V. Ward and J. A. Stanford, editors. The ecology of regulated streams. Plenum Press, New York.

Sorensen, D. L., M. M. McCarthy, E. J. Middlebrooks, and D. B. Porcella. 1977. Suspended and dissolved solids effects on freshwater biota: a review. U.S. Environmental Protection Agency, EPA Report 600/3-77-042, Washington, DC.

Starnes, L. B. 1983. Effects of surface mining on aquatic resources in North America. Fisheries 8(6):2–4.

Sullivan, K., T. E. Lisle, C. A. Dolloff, G. E. Grant, and L. M. Reid. 1987. Stream channels: the link between forests and fishes. Pages 39–96 in Salo and Cundy (1987).

Swank, W. T., and D. A. Crossley, Jr., editors. 1988. Forest hydrology and ecology at Coweeta. Springer-Verlag, New York.

Thames, J. L., editor. 1977. Reclamation and use of disturbed land in the Southwest. University of Arizona Press, Tucson.

Thirgood, J. V. 1978. Extent of disturbed land and major reclamation problems in Canada. Pages 45–68 in Schaller and Sutton (1978).

Thorne, C. R., J. C. Bathurst, and R. D. Hey, editors. 1987. Sediment transport in gravel-bed rivers. Wiley, New York.

Trautman, M. B. 1981. The fishes of Ohio. 2nd edition. Ohio State University Press, Columbus.

Vohs, P. A., I. J. Moore, and J. S. Ramsey. 1993. A critical review of the effects of turbidity on aquatic organisms in large rivers. U.S. Fish and Wildlife Service, Environmental Management Technical Center, Report 93-S002, Onalaska, Wisconsin.

Wallace, J. B., and R. W. Merritt. 1980. Filter-feeding ecology of aquatic insects. Annual Review of Entomology 25:103–132.

Wallen, I. E. 1951. The direct effect of turbidity on fishes. Bulletin of the Oklahoma Agricultural Experiment Station 48(2):1–27.

Waters, T. F. 1972. The drift of stream insects. Annual Review of Entomology 17:253–272.

Weber, P. K. 1986. The use of chemical flocculants for water clarification: a review of the literature with application to placer mining. Alaska Department of Fish and Game, Technical Report 86-4, Juneau.

White, R. J. 1975. In-stream management for wild trout. Pages 48–58 in W. King, editor. Proceedings of the wild trout management symposium. Trout Unlimited, Vienna, Virginia.

Wickett, W. P. 1958. Review of certain environmental factors affecting the production of pink and chum salmon. Journal of the Fisheries Research Board of Canada 15:1103–1126.

Wittmuss, H. D. 1987. Erosion control incorporating conservation tillage, crop rotation and structural practices. Pages 90–99 in Optimum erosion control at least cost. American Society of Agricultural Engineers, St. Joseph, Michigan.

Wolf, Ph. 1950. American problems and practice. I. Salmon which disappeared. Salmon and Trout Magazine 130:201–212.

APPENDIX II

SPECIAL BIBLIOGRAPHY ON SYMPOSIA, SEMINARS, AND CONFERENCES

ASSMR (American Society for Surface Mining and Reclamation). 1985. Symposium on the reclamation of lands disturbed by surface mining: a cornerstone for communication and understanding. ASSMR, Owensboro, Kentucky.

Blackmon, B. G., editor. 1985. Proceedings of forestry and water quality: a mid-south symposium. University of Arkansas, Department of Forest Resources, Fayetteville.

Bucek, M. F. 1981. Sedimentation ponds and their impact on water quality. Pages 345–354 in Symposium on surface mining hydrology, sedimentology and reclamation. University of Kentucky, Lexington.

Cope, O. B., editor. 1979. Proceedings of the forum—grazing riparian/stream ecosystems. Trout Unlimited, Vienna, Virginia.

Cummins, K. W., C. A. Tryon, Jr., and R. T. Hartman. 1966. Organism-substrate relationships in streams. University of Pittsburgh, Pymatuning Laboratory of Ecology, Special Publication 4, Pittsburgh, Pennsylvania.

Douglass, J. E., and O. C. Goodwin. 1980. Runoff and soil erosion from forest site preparation practices. Pages 50–74 in Forestry and water quality: what course in the '80s? U.S. Water Pollution Control Federation, Washington, DC.

Downing, W. L., editor. 1980. Proceedings, national conference on urban erosion and sediment control: institutions & technology. U.S. Environmental Protection Agency, EPA Report 905/9-80-002, Washington, DC.

Duda, A. M., D. R. Lenat, and D. Penrose. 1979. Water quality degradation in urban streams of the southeast: will non-point source controls make any difference? Pages 151–159 in International symposium on urban storm runoff. University of Kentucky, Lexington.

Eldridge, E. F., and J. N. Wilson, editors. 1959. Proceedings of the fifth symposium—Pacific Northwest on siltation—its sources and effects on the aquatic environment. U.S. Public Health Service, Water Supply and Water Pollution Control Program, Portland, Oregon.

England, R. H. 1987. Fisheries management on Georgia national forests. Pages 68–71 in J. G. Dickson and O. E. Maughn, editors. Managing southern forests for wildlife and fish, a proceedings. U.S. Forest Service General Technical Report SO-65.

Hartman, G. F., editor. 1982. Proceedings of the Carnation Creek workshop: a ten-year review. Canadian Department of Fisheries and Oceans, Nanaimo, British Columbia.

Hill, R. D. 1975. Mining impacts on trout habitat. Pages 47–57 in Proceedings of a symposium on trout habitat research and management. U.S. Forest Service, Southeastern Forest Experiment Station, Asheville, North Carolina.

Johnson, R. R., and D. A. Jones, editors. 1977. Importance, preservation, and management of riparian habitat: a symposium. U.S. Forest Service General Technical Report RM-43.

Johnson, R. R., C. D. Ziebell, D. R. Patton, P. F. Ffolliott, and R. H. Hamre, editors. 1985. Riparian ecosystems and their management: reconciling conflicting uses. U.S. Forest Service General Technical Report RM-120.

Kochenderfer, J. N., and P. J. Edwards. 1991. Effectiveness of three streamside management practices in the central Appalachians. Pages 688–700 *in* S. S. Coleman and D. G. Neary, editors. Proceedings of the sixth biennial southern silvicultural research conference. U.S. Forest Service, Southeastern Forest Experiment Station, Asheville, North Carolina.

Krumholz, L. A., editor. 1981. The warmwater streams symposium. American Fisheries Society, Southern Division, Bethesda, Maryland.

Krygier, J. T., and J. D. Hall. 1971. Proceedings of a symposium: forest land uses and stream environment. Oregon State University, Corvallis.

Marking, L. L., and T. D. Bills. 1980. Acute effects of silt and sand sedimentation on freshwater mussels. Pages 204–211 *in* J. L. Rasmussen, editor. Proceedings of the symposium on Upper Mississippi River bivalve mollusks. Upper Mississippi River Conservation Committee, Rock Island, Illinois.

Meehan, W. R., T. R. Merrell, Jr., and T. A. Hanley, editors. 1984. Fish and wildlife relationships in old-growth forests: proceedings of a symposium. American Institute of Fishery Research Biologists, Morehead City, North Carolina.

Menke, J. W., editor. 1983. Proceedings of the workshop on livestock and wildlife-fisheries relationships in the Great Basin. University of California, Division of Agricultural Sciences, Agricultural Sciences Publication 3301.

Northcote, T. G., editor. 1969. Symposium on salmon and trout in streams. H. R. MacMillan Lectures in Fisheries, University of British Columbia, Vancouver.

Oregon State University. 1988. Proceedings of the international mountain logging and Pacific Northwest skyline symposium. Oregon State University, Department of Forest Engineering, and International Union of Forestry Research Organizations, Portland.

Rice, R. M., J. S. Rothacher, and W. F. Megahan. 1972. Erosional consequences of timber harvest: an appraisal. Pages 321–329 *in* S. C. Csallany, T. G. McLaughlin, and W. D. Striffler, editors. Watersheds in transition, proceedings of a symposium. American Water Resources Association Publications, Urbana, Illinois.

Roseboom, D., R. Sauer, and D. Day. 1990. Application of a trout habitat improvement technique to Illinois warmwater streams. *In* The restoration of midwestern stream habitat. American Fisheries Society, North Central Division, Rivers and Streams Technical Committee. (Paper read at the 52nd midwest fish and wildlife conference, December 4–5, 1990, Minneapolis, Minnesota.)

Sopper, W. E., and H. W. Lull, editors. 1967. Forest hydrology: proceedings of a symposium. Pergamon Press, New York.

U.S. Bureau of Reclamation. 1948. Proceedings of the federal inter-agency sedimentation conference, 1946. U.S. Department of the Interior, Bureau of Reclamation, Denver, Colorado. (Not seen; cited in USWRC 1976.)

USDA (U.S. Department of Agriculture). 1965. Proceedings of the federal inter-agency sedimentation conference, 1963. U.S. Department of Agriculture Miscellaneous Publication 970.

USFS (U.S. Forest Service). 1975. Symposium on trout habitat research and management proceedings. U.S. Forest Service, Southeastern Forest Experiment Station, Asheville, North Carolina.

USFWS (U.S. Fish and Wildlife Service). 1977. Proceedings of a seminar: improving fish and wildlife benefits in range management. U.S. Fish and Wildlife Service FWS/OBS-77/1.

USWRC (U.S. Water Resources Council). 1976. Proceedings of the third federal inter-agency sedimentation conference, 1976. U.S. Water Resources Council, Denver, Colorado.

Vories, K. D., editor. 1976. Reclamation of western surface mined lands. Ecology Consultants, Inc., Fort Collins, Colorado.

Waddell, T. E., editor. 1986. The off-site costs of soil erosion: proceedings of a symposium. The Conservation Foundation, Washington, DC.

Wilson, J. N. 1957. Effects of turbidity and silt on aquatic life. Pages 235–239 *in* C. M. Tarzwell, editor. Transactions of a seminar on biological problems in water pollution. U.S. Public Health Service, Robert A. Taft Sanitation Engineering Center, Cincinnati, Ohio.

Wilson, J. N. 1959. The effects of erosion, silt, and other inert materials on aquatic life. Pages 269–271 *in* C. M. Tarzwell, editor. Transactions of the second seminar on biological problems in water pollution. U.S. Public Health Service, Robert A. Taft Sanitary Engineering Center, Cincinnati, Ohio.

WWRC (Washington Water Research Center). 1981. Proceedings from the conference on salmon-spawning gravel: a renewable resource in the Pacific Northwest. Washington State University, Washington Water Research Center, Report 39, Pullman.

APPENDIX III

SPECIAL BIBLIOGRAPHY OF GUIDES, MANUALS, AND HANDBOOKS

Allen, J. S., and A. C. Lopinot. 1968. Small lakes and ponds: their construction and care. Illinois Department of Conservation, Fishery Bulletin 3, Springfield.

American Farmland Trust. 1986. The economics of soil erosion: a handbook for calculating the cost of off-site damage. American Farmland Trust, Washington, DC, and Minnesota Water Conservation Board, St. Paul, Minnesota. (Available from American Farmland Trust, 1920 N Street, N.W., Suite 400, Washington, DC 20036.)

Baumann, J. S. 1990. Curbing construction site erosion. Wisconsin Natural Resources 14(4):23–28.

Bell, M. C. 1973. Fisheries handbook of engineering requirements and biological criteria. U.S. Army Corps of Engineers, Fisheries Engineering Research Program, Portland, Oregon.

Carr, W. W. 1980. A handbook for forest roadside surface erosion control. British Columbia Ministry of Forests, Land Management Report 4, Victoria.

CDFO (Canada Department of Fisheries and Oceans). 1980. Stream enhancement guide. CDFO, and British Columbia Ministry of Environment, Vancouver.

Chen, C-N. 1975. Design of sediment retention basins. Pages 285–293 *in* C. T. Haan, editor. National symposium on urban hydrology and sediment control. University of Kentucky, Lexington.

Christensen, L. A., and P. E. Norris. 1983. A comparison of tillage systems for reducing soil erosion and water pollution. U.S. Economic Research Service, Agricultural Economic Report 499, Washington, DC.

Cummins, A. B., and I. A. Given, editors. 1973a. SME mining engineering handbook, volume 1. Society of Mining Engineers of The American Institute of Mining, Metallurgical, and Petroleum Engineers, Inc., New York.

Cummins, A. B., and I. A. Given, editors. 1973b. SME mining engineering handbook, volume 2. Society of Mining Engineers of The American Institute of Mining, Metallurgical, and Petroleum Engineers, Inc., New York.

Furniss, M. J., T. D. Roelofs, and C. S. Yee. 1991. Road construction and maintenance. Pages 297–323 *in* Meehan (1991a).

Gabelhouse, D. W., Jr., R. L. Hager, and H. E. Klaassen. 1982. Producing fish and wildlife from Kansas ponds. Kansas Fish and Game Commission, Pratt.

Glover, R., M. Augustine, and M. Clar. 1978. Grading and shaping for erosion control and rapid vegetative establishment in humid regions. Pages 271–283 *in* Schaller and Sutton (1978).

Goldman, S. J., K. Jackson, and T. A. Bursztynsky. 1986. Erosion and sediment control handbook. McGraw-Hill, New York.

Gore, J. A., and F. L. Bryant. 1988. River and stream restoration. Pages 23–38 *in* J. Cairns, Jr., editor. Rehabilitating damaged ecosystems, volume I. CRC Press, Boca Raton, Florida.

Gregory, S., and L. Ashkenas. 1990. Riparian management guide, Willamette National Forest. Oregon State University, Department of Fisheries and Wildlife, Corvallis.

Hartung, R. E., and J. M. Kress. 1977. Woodlands of the northeast: erosion & sediment control guides. U.S. Soil Conservation Service and Forest Service, Upper Darby, Pennsylvania.

Harwood, G. 1979. A guide to reclaiming small tailings ponds and dumps. U.S. Forest Service General Technical Report INT-57.

Hendrickson, B. H. 1963. Conservation methods for soils of the southern Piedmont. U.S. Department of Agriculture, Agriculture Information Bulletin 269.

Hunt, R. L. 1978. Instream enhancement of trout habitat. Pages 19–27 *in* K. Hashagen, editor. Proceedings, a national symposium on wild trout management. California Trout Incorporated, San Francisco.

Hunt, R. L. 1993. Trout stream therapy. University of Wisconsin Press, Madison.

Kidd, W. J., Jr. 1963. Soil erosion control structures on skidtrails. U.S. Forest Service Research Paper INT-1.

Kochenderfer, J. N. 1970. Erosion control on logging roads in the Appalachians. U.S. Forest Service Research Paper NE-158.

Lantz, R. L. 1971. Guidelines for stream protection in logging operations. Report, Oregon State Game Commission, Portland.

Larse, R. W. 1971. Prevention and control of erosion and stream sedimentation from forest roads. Pages 76–83 *in* Krygier and Hall (1971).

Lum, F. C. H. 1977. Guides for erosion and sediment control in California. U.S. Soil Conservation Service, Miscellaneous Publication, Davis, California.

Lyons, J., and C. C. Courtney. 1990. A review of fisheries habitat improvement projects in warmwater streams, with recommendations for Wisconsin. Wisconsin Department of Natural Resources Technical Bulletin 169.

Maine Forest Service. 1991. Erosion & sediment control handbook for Maine timber harvesting operations, best management practices. Maine Forest Service, Augusta.

Nawrot, J. R., A. Woolf, and W. D. Klimstra. 1982. A guide for enhancement of fish and wildlife on abandoned mine lands in the eastern United States. U.S. Fish and Wildlife Service FWS/OBS-80/67.

NCDEHNR (North Carolina Department of Environment, Health and Natural Resources). 1989. Forestry best management practices manual. NCDEHNR, Division of Forest Resources, Raleigh.

Nielson, R. F., and H. B. Peterson. 1978. Vegetating mine tailings ponds. Pages 645–652 *in* Schaller and Sutton (1978).

NRC (National Research Council). 1982. Impacts of emerging agricultural trends on fish and wildlife habitat. National Academy Press, Washington, DC.

Packer, P. E., and E. F. Aldon. 1978. Revegetation techniques for dry regions. Pages 425–450 *in* Schaller and Sutton (1978).

Reiser, D. W., M. P. Ramey, S. Beck, T. R. Lambert, and R. E. Geary. 1989. Flushing flow recommendations for maintenance of salmonid spawning gravels in a steep, regulated stream. Regulated Rivers Research & Management 3:267–275.

Seehorn, M. E. 1992. Stream habitat improvement handbook. U.S. Forest Service Technical Publication R8-TP 16.

Swift, L. W., Jr. 1985. Forest road design to minimize erosion in the southern Appalachians. Pages 141–151 *in* Blackmon (1985).

Thornburg, A. A., and S. H. Fuchs. 1978. Plant materials and requirements for growth in dry regions. Pages 411–423 *in* Schaller and Sutton (1978).

USEPA (U.S. Environmental Protection Agency). 1973a. Methods and practices for controlling water pollution from agricultural nonpoint sources. U.S. Environmental Protection Agency, EPA Report 430/9-73-015, Washington, DC.

USEPA (U.S. Environmental Protection Agency). 1993. Guidance specifying management measures for sources of nonpoint pollution in coastal waters. U.S. Environmental Protection Agency, EPA Report 840-B-92-002, Washington, DC.

USFS (U.S. Forest Service). 1979a. User guide to vegetation: mining and reclamation in the West. U.S. Forest Service General Technical Report INT-64.

USFS (U.S. Forest Service). 1979b. User guide to soils: mining and reclamation in the West. U.S. Forest Service General Technical Report INT-68.

USFS (U.S. Forest Service). 1980. User guide to hydrology: mining and reclamation in the West. U.S. Forest Service General Technical Report INT-74.

USSCS (U.S. Soil Conservation Service). 1973. Guides for erosion & sediment control in developing areas of New Hampshire. New Hampshire Association of Conservation Districts, and the North Country Resource Conservation and Development Project, Durham.

USSCS (U.S. Soil Conservation Service). 1982. Ponds—planning, design, construction. U.S. Soil Conservation Service, Washington, DC.

USSCS (U.S. Soil Conservation Service). 1983. National engineering handbook, section 3, sedimentation. U.S. Department of Agriculture, Washington, DC.

WDNR (Wisconsin Department of Natural Resources). 1992. Wisconsin construction site best management practice handbook (revised). Wisconsin Department of Natural Resources, Bureau of Water Resources Management, Madison.

White, R. J., and O. M. Brynildson. 1967. Guidelines for management of trout stream habitat in Wisconsin. Wisconsin Department of Natural Resources Technical Bulletin 39.

Woodward–Clyde Consultants. 1980b. Gravel removal guidelines manual for Arctic and Subarctic floodplains. U.S. Fish and Wildlife Service FWS/OBS-80/09.

Yee, C. S., and T. D. Roelofs. 1980. Planning forest roads to protect salmonid habitat, part 4. Influence of forest and rangeland management on anadromous fish habitat in western North America. U.S. Forest Service General Technical Report PNW-109.

APPENDIX IV

THESES AND DISSERTATIONS CITED IN THE SEDIMENT LITERATURE

Angle, L. 1987. Effects of sediment addition on the drift of aquatic macroinvertebrates in Nine Mile Creek, Nebraska. Master's thesis. University of Nebraska, Lincoln.

Bachman, R. W. 1958. The ecology of four north Idaho trout streams with reference to the influence of forest road construction. Master's thesis. University of Idaho, Moscow.

Berg, L. 1983. Effects of short-term exposure to suspended sediments on the behavior of juvenile coho salmon. Master's thesis. University of British Columbia, Vancouver.

Bianchi, D. R. 1953. The effects of sedimentation on egg survival of rainbow trout and cutthroat trout. Master's thesis. Montana State College, Bozeman.

Crouse, M. R. 1982. Effects of fine sediments and substrate size on growth of juvenile coho salmon in laboratory streams. Master's thesis. Oregon State University, Corvallis.

Deol, U. T. S. 1967. The effect of inorganic pollution on macroinvertebrate populations of Deer Creek. Master's thesis. DePauw University, Greencastle, Indiana.

Dillard, S. M. 1992. Investigation of fine sediment effects on the benthic macroinvertebrates of a southern Blue Ridge escarpment stream: implications for assessing impacts from erosion. Master's thesis. Clemson University, Clemson, South Carolina.

Drewes, H. G. 1984. Factors affecting water quality and macroinvertebrate distribution within a small Black Hills stream. Master's thesis. South Dakota State University, Brookings.

Garman, G. C. 1984. Initial effects of deforestation on aquatic community structure and function of the East Branch Piscataquis River, Maine. Doctoral dissertation. University of Maine, Orono.

Golladay, S. W. 1988. The effects of forest disturbance on stream stability. Doctoral dissertation. Virginia Polytechnic Institute and State University, Blacksburg.

Gregory, S. V. 1979. Primary production in streams of the Cascade Mountains. Doctoral dissertation. Oregon State University, Corvallis.

Hains, J. J. 1981. The response of stream flora to watershed perturbations. Master's thesis. Clemson University, Clemson, South Carolina.

Hausle, D. A. 1973. Factors influencing embryonic survival and emergence of brook trout (*Salvelinus fontinalis*). Master's thesis. University of Wisconsin, Stevens Point.

Hess, L. J. 1969. The effects of logging road construction on insect drop into a small coastal stream. Master's thesis. Humboldt State College, Arcata, California.

Kelley, T. L. 1974. Methodology for in-stream rehabilitation of a silted stream. Master's thesis. University of Idaho, Moscow.

King, D. L. 1964. An ecological and pollution-related study of a warm-water stream. Doctoral dissertation. Michigan State University, East Lansing.

Klamt, R. R. 1976. The effects of coarse granite sand on the distribution and abundance of salmonids in the Central Idaho Batholith. Master's thesis. University of Idaho, Moscow.

Konopacky, R. C. 1984. Sedimentation and productivity in salmonid streams. Doctoral dissertation. University of Idaho, Moscow.

Koski, K. V. 1966. The survival of coho salmon (*Oncorhynchus kisutch*) from egg deposition to emergence in three Oregon coastal streams. Master's thesis. Oregon State University, Corvallis.

Koski, K. V. 1975. The survival and fitness of two stocks of chum salmon (*Oncorhynchus keta*) from egg deposition to emergence in a controlled-stream environment at Big Beef Creek. Doctoral dissertation. University of Idaho, Moscow.

Lacy, G. F. 1982. First year effects of salvage clearcut logging upon stream populations of wild brook and brown trout in the northcentral highland region of Pennsylvania. Master's thesis. Pennsylvania State University, College Park.

MacFarlane, M. B. 1978. Effects of silt on benthic macroinvertebrates in the Redwood River, Redwood County, Minnesota. Doctoral dissertation. University of Minnesota, St. Paul.

Marsh, P. C. 1979. The effects of silt and turbidity from agricultural drainage on benthic invertebrates in streams of southwestern Minnesota. Doctoral dissertation. University of Minnesota, St. Paul.

Martin, D. J. 1976. The effects of sediment and organic detritus on the production of benthic macro-invertebrates in four tributary streams of the Clearwater River, Washington. Master's thesis. University of Washington, Seattle.

Matter, W. H. 1979. The status and seasonal dynamics of fish and benthic invertebrate populations in relation to organic and inorganic material inputs and surface mining impacts in three Virginia headwater streams. Doctoral dissertation. Virginia Polytechnic Institute and State University, Blacksburg.

McClelland, W. T. 1972. The effect of introduced sediment on the ecology and behavior of stream insects. Doctoral dissertation. University of Idaho, Moscow.

McCuddin, M. E. 1977. Survival of salmon and trout embryos and fry in gravel-sand mixtures. Master's thesis. University of Idaho, Moscow.

Newbold, J. D. 1978. The use of benthic macroinvertebrates as indicators of logging impact on streams with an evaluation of buffer strip effectiveness. Doctoral dissertation. University of California, Berkeley.

Noggle, C. C. 1978. Behavioral, physiological, and lethal effects of suspended sediment on juvenile salmonids. Master's thesis. University of Washington, Seattle.

Prather, K. V. 1971. The effects of stream substrates on the distribution and abundance of aquatic insects. Master's thesis. University of Idaho, Moscow.

Reis, P. A. 1969. Effects of inorganic limestone sediment and suspension on the eggs and fry of *Brachydanio rerio*. Master's thesis. DePauw University, Greencastle, Indiana.

Ringler, N. H. 1970. Effects of logging on the spawning bed environment in two Oregon coastal streams. Master's thesis. Oregon State University, Corvallis.

Rogers, B. A. 1969. Tolerance levels of four species of estuarine fishes to suspended mineral solids. Master's thesis. University of Rhode Island, Kingston.

Sandine, M. F. 1974. Natural and simulated insect-substrate relationships in Idaho Batholith streams. Master's thesis. University of Idaho, Moscow.

Sigler, J. W. 1981. Effects of chronic turbidity on feeding, growth and social behavior of steelhead trout and coho salmon. Doctoral dissertation. University of Idaho, Moscow.

Simmons, R. C. 1984. Effects of placer mining sedimentation on Arctic grayling of interior Alaska. Master's thesis. University of Alaska, Fairbanks.

Smith, D. W. 1978. Tolerance of juvenile chum salmon (*Oncorhynchus keta*) to suspended sediments. Master's thesis. University of Washington, Seattle.

Sowden, T. K. 1983. The influence of groundwater and spawning-gravel composition on the reproductive success of rainbow trout in an eastern Lake Erie tributary. Master's thesis. University of Waterloo, Waterloo, Ontario.

Steuhrenberg, L. L. 1975. The effects of granitic sand on the distribution and abundance of salmonids in Idaho streams. Master's thesis. University of Idaho, Moscow.

Tagart, J. V. 1976. The survival from egg deposition to emergence of coho salmon in the Clearwater River, Jefferson County, Washington. Master's thesis. University of Washington, Seattle.

Vosick, D. J. 1988. A critical evaluation of techniques used to calculate the economic cost of off-site damage caused by soil erosion in two watersheds in Minnesota. Master's thesis. University of Minnesota, St. Paul.

Wagener, S. M. 1984. Effects of placer mining on stream macroinvertebrates of interior Alaska. Master's thesis. University of Alaska, Fairbanks.

Wolcott, L. T. 1990. Coal waste deposition and the distribution of freshwater mussels in the Powell River, Virginia. Master's thesis. Virginia Polytechnic Institute and State University, Blacksburg.

Wustenberg, D. W. 1954. A preliminary survey of the influences of controlled logging on a trout stream in the H. J. Andrews Experimental Forest, Oregon. Master's thesis. Oregon State College, Corvallis.

INDEX

When a subject appears two or more times on the same page, but in different discussions, the corresponding page number is repeated according to the number of subject appearances. When a subject is scattered over a range of pages, *"passim"* follows the page numbers.

sedimentation in streams of 67, 103;
  Plate 9
suspended sediment levels in 60
trout streams in 102, 103, 103, 104, 107–
  108, 109, 170
Aquatic Life Advisory Committee (Ohio
  River) 65
Arctic floodplains, gravel mining in 41–42
area mining, as type of strip-mining for
  coal 39
arid regions, fisheries in 24
Arizona, sediment studies in 50, 134
Arkansas, sediment studies in 134
Army, U.S. Department of the, formed
  committee to study sediment 4
assimilative capacity of streams 1, 39, 169
avalanches, associated with logging
  roads 29
  see also landslides; mass soil movements
avoidance of sediment, by fish 81–83, 174

Baltimore, Maryland, urban development
  in 42, 172
barite mining 40
bass, largemouth 117
bass, smallmouth 8, 74, 83, 117
bass, spotted 117
bass, striped 110
Bear Creek, Mississippi 72
Bear River, California 37
bed load, of sediment
  and experimental sand additions to 115
  interdiction of, by sediment traps 152,
    155, 178
  and log yarding 62
  and patch-cut logging 133
beetles *see* Coleoptera
benthic fishes 72
benthos
  abundance 62, 62, 63–65, 73
  biomass (standing stock) 60, 75, 125
  diversity 150
  habitat 125
  in interstitial space 72
  populations 59, 60, 78
  production 58, 61, 61, 72, 75
  recolonization of, by drift 48, 49
  recovery 70
  reduction of, by sediment 40, 42, 48, 70,
    78, 115, 129
  and turbidity 58
  see also bottom fauna; insect production;
    invertebrate production;
    invertebrates, aquatic

bentonite, sealing a pond bottom with 53
best management practice (BMP) to reduce
  sediment
  in forest management 131
  in urban development 44
Bighorn River, Montana, tailwater fishery
  in 154
Big Rib River, Wisconsin 41
biological monitoring, of invertebrates 68
bluegill 117
Bluewater Creek, Montana 20
bottom fauna, and sediment 65, 65, 80, 113
  see also benthos; insect; insects, aquatic;
    invertebrate; invertebrates, aquatic
boulders, as invertebrate substrate 63, 68
Brazos River, Texas 58, 73, 117
bridge construction, as source of
  sediment 17, 30, 35, 35, 49, 77, 111,
    115, 115, 133, 143, 172, 173, 176;
    Plate 48
British Columbia
  Flathead Lake tributaries in 102
  gravel-cleaning operations in 167
  pipeline crossing studies in 70
  Queen Charlotte Islands of 34
  salmonid reproduction studies in 72, 96
  sediment studies in 33, 34, 66, 72, 72, 87,
    114, 115, 134
  see also Carnation Creek Watershed
    Project
British Columbia Ministry of Forests and
  Lands 136
broad-based dip, for sediment control on
  logging roads 131, 134, 176;
  Figure 131; Plate 132
Buckhorn Dam, on Grass Valley Creek,
  California 154
buffer strips
  and agriculture 20, 149, 150, 178
  in Conservation Reserve Program 150
  on floodplains 124, 152, 178
  and forestry 60, 29, 35, 35, 72, 97, 99, 134,
    149, 150, 151, 152, 177
  and livestock grazing 149, 152
  as means of sediment interdiction 149–
    152, 177
  and mining 152, 178
  around ponds 153
  in riparian zone 149, 150, 151, 152, 177,
    177–178
  and sand and gravel mining 143
  and sedimentation 72, 151
  stability of 151
  and streambank erosion 72, 151, 151

sediment and stream fishes in 26
sediment control measures in 126
streams in 89, 109
studies on spawning gravels in 33
U.S. Forest Service Experiment Station, at
    Portland, Oregon 33
Pacific Southwest Forest Service
    Experiment Station, at Berkeley,
    California 33
panning, for placer gold deposits 37
particle size, of sediments
    and benthic invertebrates 61, 86
    in salmonid redds 100, 103, 104, 106–
        107, 109
    in spawning gravels 103
patch-cutting, of timber
    in experimental timber harvest 97, 131,
        133
    in sediment control 134
    as source of sediment 24, 29–35 *passim*
pelagophils, fish reproductive guild 110
pentachlorophenol, as toxicant to fish 84
perch, white 110
periphyton, and sediment 5, 13, 53, 56, 57,
    69, 74
permeability, of salmonid redd gravels 30,
    95, 98, 103
pesticides, as co-pollutant with
    sediment 20, 21
Pheasant Branch, Wisconsin 43
phosphate mining 40
photosynthesis 14, 41, 53, 55, 56, 57, 57, 63,
    77, 99, 173
    see also primary production
Pigeon River, Michigan 104
pine, loblolly 134
pine, white 46
Pine River, Michigan 45–46
pipeline crossing of streams, as source of
    sediment 17, 51, 70, 77, 173
Piscataquis River watershed, Maine 28, 151
Pittsburgh, University of 63
placer mining, as source of sediment 36–37
    passim, 42, 58, 73, 80, 82–83 *passim*,
    136, 171
planting, of fish eggs 87, 88, 88, 102, 103,
    103, 107
Platts, W. S., biologist 32
Plecoptera (stoneflies) 7, 20, 20, 63, 66, 76,
    173–174
    see also EPT
plunge pool, as stream fish habitat 162,
    162, 163, 164; Plate 112
ponds

constructed by excavation 153
constructed by impoundment 153
for fish production 153, 153–154
as sediment traps 153
tailings 40, 140, 140–141, 152, 171, 176
see also farm ponds
pool cover
    scouring of, by flushing sediment 157
    scouring of, by habitat alteration 157,
        162, 163–164, 179
    for stream salmonids 111, 111, 113, 114–
        115, 118, 175
Poplar Creek, Michigan 155
Potomac River, Virginia 47–48
preventive measures of sediment control
    in agriculture, livestock grazing 124–
        126, 176
    in agriculture, row-crops 119–124, 176
    to eliminate source of sediment 119–149,
        157, 176–177
    in forestry 126–135, 176
    in mining 135–143, 176–177
    in streambank erosion 146–149, 177
    in urban development 143–146, 177
primary production
    affected by deposited sediment 57–58
    affected by light from canopy
        removal 57, 72, 73, 75, 77
    affected by suspended sediment 53, 55–
        58, 173
    and agriculture 19
    as element of stream communities 5, 7
    and land use 20
    and mining waste 37
    and phosphorus 72
    see also photosynthesis
Prince Edward Island, siltation of stream
    in 114
production
    benthos *see* benthos production
    biological, and sediment 4
    fish *see* fish production
    invertebrate *see* invertebrate production
    salmonid *see* salmonid production
    trout *see* trout production
Public Health Service, U.S.
    "Fifth Symposium," on siltation 5
    initiated research on water pollution 5
    Robert A. Taft Sanitary Engineering
        Center 63

quarrying, and sediment production 42
Queen Charlotte Islands, British Columbia,
    logging in 35, 46